DIY SOLAR POWER FOR BEGINNERS

A TECHNICAL GUIDE ON HOW TO DESIGN, INSTALL, AND MAINTAIN GRID-TIED AND OFF-GRID SOLAR POWER SYSTEMS FOR YOUR HOME

DIY SOURCE

DIY SOLAR POWER FOR BEGINNERS

© Copyright 2021 - All rights reserved.

It is not legal to reproduce, duplicate, or transmit any part of this document in either electronic means or in printed format. Recording of this publication is strictly prohibited and any storage of this document is not allowed unless with written permission from the publisher except for the use of brief quotations in a book review.

Disclaimer

Certain home improvement projects are inherently dangerous, and even the most benign tool can cause serious injury or death if not used properly. ALWAYS READ AND FOLLOW INSTRUCTION MANUALS AND SAFETY WARNINGS. You must be particularly careful when dealing with electricity—always use common sense.

Any advice, guidance, or other information provided on the electrical-online.com website or within any of our publications cannot completely anticipate your situation. If you are at all unsure about completing any aspect of this or other home wiring projects, consult a qualified electrical contractor to perform the service(s) for you.

ALWAYS follow electrical code requirements specific to your area. Before undertaking any electrical project, contact your local electrical authority and insurance company to ensure that you comply with all policies, warranties and regulations concerning this work.

DIY SOLAR POWER FOR BEGINNERS

Special Bonus!

Want this Bonus Book for FREE?

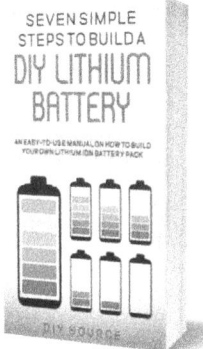

Get FREE, unlimited access to it and all of our new books by joining the Fan Base!

  *Scan W/ Your Camera to Join*

DIY SOLAR POWER FOR BEGINNERS

Table of Contents

INTRODUCTION — *13*

CHAPTER ONE: DIY Solar Power System — *15*

  Types of Solar Power Systems — 17

  DIY Installation of a Solar Power System — 21

  Steps to Install a DIY PV System — 25

CHAPTER TWO: Basic Electricity Rules, Formulas, and Circuits — *27*

  Primary Factors of Electricity — 27

  Ohm's Law — 32

  Watt's Law — 34

  Kirchhoff's Law — 38

  Electrical Circuits — 40

  Types of Connection (series, parallel) — 45

CHAPTER THREE: Essential Tools and Equipment — *53*

  Safety Tools — 54

  Power Tools — 56

  Wiring Tools and Equipment — 61

  Solar Racking Equipment — 68

  Battery & Maintenance Tools — 70

CHAPTER FOUR: Types of PV Systems and Components ___ 73

Grid-Tied Solar Power System ___ 73

Components of a Grid-tied System ___ 76

Grid-Tied Solar Power System with Battery Backup (Hybrid Solar System) ___ 78

Components of a Hybrid System ___ 79

All-in-One Solar Power Systems ___ 83

Off-grid Solar Power Systems (Standalone Solar Power Systems) ___ 84

Components of an Off-Grid Solar Power system ___ 86

CHAPTER FIVE: Solar Panels (PV Modules) ___ 91

Types of PV Modules ___ 92

Solar Panel Specification Sheet ___ 94

PV Module and Shading Effects ___ 104

Shading of the PV Array/String ___ 106

Tilt and Orientation of the PV Array ___ 112

PV Sizing for Grid-Tied Systems ___ 123

How to Size Hybrid Solar Systems ___ 129

How to Size an Off-Grid PV System ___ 132

Roof Sizing ___ 139

Choosing Suitable Solar Panels ___ 142

Pros and Cons of Roof Mount PV System ___ 147

    Pros and Cons of Ground Mount PV System 148
    Mounting Systems _____ 149
    Roof Mounting Installation Process _____ 153

CHAPTER SIX: Solar Charge Controllers ____ 159
    Why do we need a charge controller? _____ 160
    Charge Controllers and Stages of Charging 161
    Charge Controller Technologies _____ 163
    How to Size a PWM Controller _____ 166
    Size an MPPT Charge Controller _____ 168
    Choose the Right Solar Controller: PWM versus MPPT _____ 171
    Programming Solar Charge Controllers ___ 172
    Connect the Charge Controller _____ 176

CHAPTER SEVEN: Solar Battery Bank _____ 179
    What Is a Deep Cycle Battery? _____ 179
    Types of Batteries _____ 180
    Types of Lead-Acid Batteries _____ 181
    Lead-Acid Battery Features and Terminologies _____ 182
    Monitoring a flooded lead-acid battery ____ 184
    Temperature Effect on Lead-Acid Batteries 187
    VRLA vs. FLA Battery Charging and Maintenance _____ 190
    Lithium-Ion Battery _____ 191

Comparing Lithium-Ion and Lead-Acid Battery 192

A SHORT MESSAGE FROM Error! Bookmark not defined.

"DIY SOURCE BOOKS" Error! Bookmark not defined.

About DIY SOURCE BOOKS 197

CHAPTER EIGHT: Solar Inverters 199

    Inverter's Function 199

    Common Features of Inverters 201

    Voltage and Power Ratings (Wattage) 202

    Off-Grid Inverters 206

    Grid-Tied Inverters 207

    Hybrid Inverters 208

    AC-Coupled and DC-Coupled Configurations 209

    How to Size an Inverter 214

CHAPTER NINE: Conductors and Connectors 219

    Composition of Wires and Cables 219

    Wire Color Codes and Solar Application 223

    Basic Principles of Wire Sizing 226

    Sizing Conductors for PV Circuits 235

    PV Source Wire 236

    PV Output Wires 241

    Battery to the Inverter Wires 245

Inverter Output Cable _____ 248

PV Module Connectors _____ 249

Wiring Solar Panels (stringing) _____ 250

CHAPTER TEN: Connecting Overcurrent Protection Devices, Wrapping It Up, and Troubleshooting _____ *259*

DC and AC Load Centers _____ 260

Basic Rules for OCP Device Sizing _____ 261

How to Crimp _____ 266

Final Checklist and Troubleshooting _____ 268

Common Mistakes Solar Installers Make __ 271

Final Words _____ *275*

References _____ *281*

DIY SOLAR POWER FOR BEGINNERS

INTRODUCTION

The desire to use green sources of energy and be energy independent has led to the increased use of PV solar power. This comprehensive guide covers everything from designing and assembling rooftop racking systems or ground-mount structures to setting up an electrical circuit connection for your household devices.

Our guide is beginner-friendly as it provides you with easy steps on how to design and install a solar power system from scratch. People who haven't mastered the installation process can follow this guide to install a solar array to maximize their power output.

In this DIY solar power guide, you will learn:

• The critical aspects you need to know before installing a solar system.

• Basic electrical rules and circuits.

• Solar installation tools and required equipment.

• The different types of solar setups to help you choose the best type of solar system to install, based on your energy needs.

• How to determine the size of your solar array and some examples of cost-effective solar panels.

• How to size your roof and install a racking system on your rooftops or on the ground.

- How to maximize solar output based on your location, tilt angle, azimuth angle, and panel orientation.

- Different types of charge controllers and how to size them.

- Types of battery systems, AC-coupled versus DC-coupled systems.

- Sizing battery for your solar system.

- How to size and connect a solar inverter to your system.

- Wire and circuit breaker sizing and different types of connectors for your solar setup.

- Mistakes to avoid when installing a solar power system.

- How to troubleshoot your solar power system.

These simple steps will guide you on how to choose, design, and install a reliable and efficient solar power system that meets your energy demands and cuts down on your utility bills. You need this Book!

CHAPTER ONE: DIY Solar Power System

The increasing cost of electricity has left many people asking themselves, "Should I go solar?" If you look around your town or city, you'll probably see more and more solar panels. People are welcoming the idea of living off the grid while still having a reliable power source.

A solar power system provides you with sustainable clean energy that can save you a considerable amount of money per year. Solar-powered systems utilize a renewable source of energy with zero emissions of greenhouse gasses. This type of energy is classified as green energy because of its environmental and economic benefits.

What Is a Solar Power System?

It is a collection of solar panels that converts energy harnessed from sunlight into electricity. Solar panels use photovoltaic (PV) technology that converts the energy harnessed from the sun into an electric current. Solar setup generates electricity to use for lighting and heating buildings, cooling, and more.

You can have a single solar panel that provides electricity for charging your batteries or lighting. Or you can scale up to a large array of panels that provide energy for your entire home.

A majority of homeowners in developed countries have installed or are planning to add a solar power system into their grid to supplement the conventional sources of energy. Even in less developed nations, solar energy is believed to be the solution to the world's impending energy crisis.

If you're looking forward to designing and installing a solar setup in your home, our guide will walk you through the whole process of setting up a DIY solar power system. In the end, you not only can supplement your energy sources, but also have an efficient solar power system that meets your energy needs.

Want to find out how? Keep reading!

How Does The Solar Power System Work?

Tiny packets of energy (photons) from the sun strike the earth's surface, generating enough solar energy that can satisfy a great portion of global power consumption. When the photons hit the solar cells in a solar panel, they create free electrons that generate the flow of electricity.

Solar setups utilize photovoltaic (PV) technology to capture solar radiation and turn it into a useful form of energy. Solar panels are made of varying numbers of cells that are used to generate electricity. The cells are made of a semiconductor material called silicon, which absorbs the sun ray's energy. This results in the creation of electrically charged particles that move in response to the internal electric field inside the PV cell and convert the solar radiation into electricity.

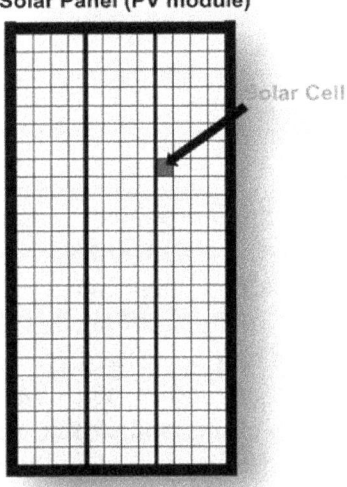

Silicon in PV cells is arranged in two layers; the positively charged layer and the negatively charged layer. When sunlight hits the cells, the electric field between the two layers causes electricity to flow and generate a direct current (DC). You can draw this current for external use in your home.

Though you don't always need bright sunlight for your solar panels to work, the intensity of the light determines the amount of electricity that can be generated. As the generated electricity exceeds your needs, you can invest in batteries to store power for use at night or during cloudy days.

Types of Solar Power Systems

A solar power system's integration into electrical grids and homes can be classified under three methods:

- Grid-Tied
- Off-Grid
- Hybrid

Grid-Tied Solar Power System

A grid-tied system requires a connection to a network or a utility power grid in order to supply energy. There are no batteries connected to the system and any excess energy that is generated is sent back to the network. If the energy generated by your PV system is more than your power consumption, then the excess electricity is directed to the grid through the meter. This specific meter records the amount of kilowatt hour (KWh) sent to the grid. Your utility company will record this figure in every billing cycle and pay you accordingly.

Off-Grid Solar Power System

An off-grid system doesn't require a connection to the electricity grid. You only need to connect to the battery to store the excess electricity. The battery capacity will enable you to store enough electricity to be used at nights and during cloudy days. The power generated can meet residential, industrial, and commercial user needs even during the winter months.

Hybrid Solar Power System

A hybrid system is a mix of both grid-tied and off-grid systems. When the grid is available, they work as on-grid solar systems; otherwise, they perform like the off-grid system. They generate electricity similar to a conventional grid-tied solar system.

The lithium or lead-acid battery connected acts as a backup power supply. If you want to have a continuous power supply in case of a power outage, you should consider a hybrid PV system.

How Long Does It Take to Pay Off (Grid-Tied and Hybrid Systems)?

When you go solar, you reduce your monthly electricity bill paid to the utility company. If you have a grid-tied solar system, you can export the excess solar energy to the utility company and receive payment on a quarterly basis based on how much electricity you send to them.

When your solar panels don't generate enough electricity, you can import more electricity from the utility company.

This will reduce your electricity bill compared to when you're solely dependent on the utility company.

Investing in solar energy is a decent decision you can make to reduce your expenditure and keep the environment safe. Though the initial cost of setting up a solar system is high, over time, the system will pay you back. The state and federal tax incentives accelerate the payoff schedule. If you install a grid-tied solar system, it can pay itself off within three to six years if you install it yourself, and around five to nine years should you hire a professional contractor.

In addition, solar panels come with a 25-year warranty and the energy generated beyond the initial payback period is considered a return on your investment (ROI).

 The payback period is how long it takes you to recover from your initial investment installing a solar system. You can easily use this online solar ROI calculator to calculate your payback period. Alternatively, you can use the solar payback formula to determine how long it will take to pay off. You can start by calculating the total cost of installing the solar system after deducting the incentives. Then you can compare this figure against the cost of electricity from the utility company.

*Payback Period = (Total Installation Cost - Value of Incentives) / (Cost of Electricity per Unit * Annual Electricity Consumption)*

Your total installation costs include the cost of solar equipment, permit, installation cost, contractor wages, and any other costs associated with the project. Value of incentives represents a reward such as tax credits you get for installing solar power. The solar investment tax credit is usually 26% of federal taxes and you're entitled to claim it. You can check these credits by scanning the code below.

 Check other local or state incentives in your country when calculating the payback period. The cost of electricity is the billing rate per KWh of electricity.

You can get information about the cost of electricity from your utility provider. Electricity consumption or usage is usually printed in your electricity bill.

Annual electricity consumption is the total monthly usage multiplied by twelve to get your annual consumption. Alternatively, you can gather full-year monthly bills from the utility company and get an accurate amount.

Decreasing Costs of Solar Power

Due to the development of cost-effective and efficient power systems, energy costs have dropped. According to the International Renewable Energy Agency (IRENA) 2019 report, energy costs were reduced by 82%, while that of photovoltaic solar energy dropped by 47%.

The reduction in costs was a result of improved technology, changes in the economies of scale, and the competitiveness of the supply chain. This led to the rise of the global capacity of solar from 40 GW (gigawatt) to 580 GW between 2010 and 2019, according to IRENA. Today, Photovoltaic (PV) projects aimed at utilizing renewable green sources of energy are increasingly becoming less expensive.

DIY Installation of a Solar Power System

The use of solar power in residential areas has highly contributed to the reduced installation costs. You can use the DIY option to keep the installation cost within your budget. So can you install a solar power system by yourself?

Yes, if you live in the US, for example, you can definitely install your own solar system and save money; but if, for instance, you live in Australia, you have to work with an accredited electrician and comply with the set standards and legislatures.

Solar system installation may seem highly complex, but it is doable! You only need to have basic electrical knowledge and you will be ready to do it yourself. However, you have to take into account a number of other considerations before installing a solar power system in your home. To start with, you have to estimate your electricity needs and whether you will use solar power all year round. Different homes have different electricity needs. In some homes, electricity is needed to power lights and charge cell phones, while, in other homes, electricity is needed to run a television, refrigerator, and other items.

Factors to Consider before Setting up a PV System

Before buying any solar components and installing them, there are a number of factors you have to consider. Since solar system installation is a very large investment, you have to be accurate in your calculations and your choices.

Knowing these key facts will make it easy for you to select the right products for your solar setup. Some of these factors are as follows:

1. How long will you live in your house?

Though you may notice an immediate drop in your electricity bills after installing a solar power system, it will take at least five to six years for your solar investment to pay off. If you intend to live in the same house for the long term, then your investment in solar power is worth every cent; otherwise, the cost of solar powering your house will be much higher.

2. How much energy do you need to power your house?

Another important fact you have to consider is the amount of energy you need to run all the appliances in your home. Of course, different households have different electricity needs based on the type of devices they're using and the number of people living in that house.

If you have a lower consumption level, you can go for fewer solar panels. So it is important to know your average power consumption to estimate how many solar panels are sufficient for your energy needs.

Generally, the size of your solar power system depends on insolation (sun exposure) and how much energy you need.

3. Cost of solar system

The cost of solar equipment varies from one manufacturer to another, making it difficult to know which equipment is appropriate and not overpriced. Though the prices of solar panels have significantly reduced, other equipment and installation expenses might be high. Altogether, it may take you five to six years to offset the cost of deploying a PV system.

4. Roof type

Although you can install solar panels on almost every rooftop, some of them require more effort and have to incur extra costs to install the solar panels. For instance, it is very difficult to install solar panels on regular wood shingle rooftops.

5. Location

The amount of energy generated depends on where you live and the amount of direct sunlight exposure to your panels. If you live in areas where there are tall trees and buildings, it can affect sunlight exposure to your panels.

If you live in the US and in a state like California, Texas, or Arizona, you may experience high energy output from your solar system since these areas receive full sunlight for long hours; however, if you live in areas with less direct sunlight like Montana or Minnesota, you may still need to rely on the grid, while utilizing solar energy as much as possible by equipping a more extensive or efficient PV system with larger, extra solar panels.

6. Permits

Before you install solar panels in your home, you have to file for an electrical permit, a structural or building permit, and a dedicated solar photovoltaic permit. Consult a local professional to know all the legal issues and zoning laws within your country.

You should also obtain relevant information in regard to building permits to avoid the risk of having to re-do the installation of the whole system or be fined. This is especially true regarding ground-mounting solar panels.

Mostly, building permits are issued at the local level, so it is important to follow all the state rules and regulations that apply to your municipality.

Steps to Install a DIY PV System

1. Design and size your solar power system based on your power needs.

2. Buy solar power equipment.

3. Mount your solar panels either on the rooftop or in your garden.

4. Connect your solar panels to the charge controller and then to the battery (in battery-backed systems).

5. Install solar inverters and set up stands for your inverter and battery.

6. Install a smart meter/net metering (for grid-tied and hybrid systems).

7. Complete your power system by connecting it to the main electrical board.

8. Request your local utility company to give you Permission to Operate (PTO) on the net metering system and connect to the grid (if applicable).

Following these steps will help you set up your solar power unit. The financial return can be reaped later since solar energy is a green source of energy and also cost-effective.

To learn more about how to set up your own system, stick around and keep reading.

Before we can go into detail on how to go about the installation process, let's look at basic electricity rules and formulas in the next chapter.

Chapter Summary

Solar energy is a renewable, green energy source generated from the sun that is converted to electricity by photovoltaic (PV) technology that harnesses this energy and delivers it to different electrical appliances.

Solar energy is also cost-effective; deciding to go solar will reduce your monthly electricity bill from the utility supplier and can lead to utility independence.

There are different types of solar power systems you can choose based on your needs and installation process. Deciding on whether to be grid-tied or not is based on clear-cut benefits of grid-tied solar systems and individual options. The majority of homeowners prefer grid-tied connections. However, there are those who prefer off-grid or hybrid solar power systems.

CHAPTER TWO: Basic Electricity Rules, Formulas, and Circuits

Electricity uses are everywhere, hence the need to have basic knowledge on laws that govern how electricity operates. To understand these laws, you need to understand the key foundation of basic electrical concepts such as voltage, resistance, current, Ohm's law, circuit theory, and others. Let's start by defining what electricity is.

Electricity is the flow of an electric charge. An electric charge can either be positive or negative, and its movement creates an electric field. The charge is generated from primary sources (natural sources of energy such as sunlight) and secondary sources.

No matter how the charge is created, the flow of this charge results in an electric current. In this tutorial, you will learn the basic electrical concepts that help you better utilize them while designing and installing your PV system.

Primary Factors of Electricity

The fundamental forces of electricity that control all electrical circuits include:

- ➢ Voltage
- ➢ Current flow
- ➢ Resistance (Impedance)

Voltage

Voltage or potential difference is an electromotive force that pushes the current to flow through an electrical circuit. It is measured in *volts* and denoted by the letter "V". Scientifically, a volt is defined as the electromotive force (E) required forcing a single ampere of current to flow through a resistance of one ohm.

Let's compare voltage to water pressure in a water hose. If the pressure is very high, the water will flow through the system much faster. Similarly, if the voltage (electrical pressure) is high, then electricity will flow through the circuit at a faster rate.

Current

As explained previously, the flow of electrical charge results in an electric current. The rate of flow of the current is measured in *amperes (A)*, and it is represented by the letter "I". The scientific definition of the current is the flow of $6.25 * 10^{23}$ electrons per second.

Considering the same water hose analogy, just as voltage resembles the water pressure, current can be compared to the amount of water passing through the hose pipe. And just like the rate of flow of water through the pipe, current represents the rate of flow of electrons through the conductor. For example, the number of electrons flowing through a circuit with 12A current will be three times as those flowing through a circuit with a current of 4A.

The current flowing through a circuit can either be direct current (DC) or alternating current (AC). DC has a constant voltage polarity that allows it to flow in only one direction. By contrast, AC flows in both directions along with its voltage polarity. The diagram below shows the difference between two types of electric power.

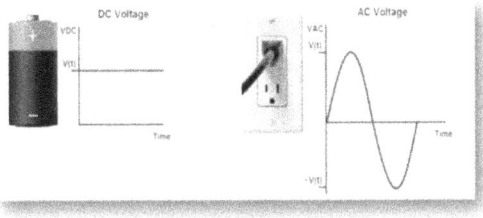

AC Power

Alternating current (AC) power is the standard electricity emitted from the power outlets. The flow of electric charge periodically changes from either positive (upward) to negative (downward) direction. The movement of electrons results in the formation of sinusoidal AC waves.

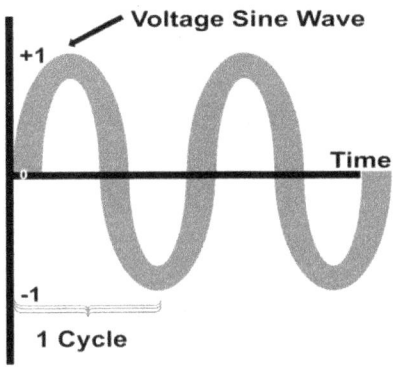

AC power produced by an alternator switches its polarity; this is due to the movement of the stationary coil in relation to magnetic flux.

As depicted in the diagram, both AC current and voltage follow a particular sinusoidal pattern. Depending on the type of the load of the electrical device used, these waveforms may be in phase or out of phase in relation to each other.

Sinusoidal AC waves vary from -1 (located under the horizontal line) to +1 (located above the line). For example, a sine of 90 degrees is considered 1 while a sine of 0 degrees is equal to 0. The voltage and current waveforms flowing on an oscilloscope have sine waves that overlap each other. The oscilloscope is the device used to measure the AC sine waves.

Each complete sine wave is called a cycle, which consists of two negative and positive peaks (+1 and -1) each located between two points of zero (illustrated in the diagram). Frequency is the measure used to describe the alternating rate of both voltage and current. The unit used to measure the wave's frequency is called hertz. The standard power frequency in the US is 60 hertz, which is equal to 60 complete sine waves (cycles) in one second.

DC Power

Direct current (DC) is a linear electrical current that moves in a straight line. DC power is drawn from batteries, fuel cells, and solar cells. You can also obtain DC from AC power by using an inverter or a rectifier that converts AC to DC power.

Most electronic devices use DC power from batteries because batteries offer consistent voltage. Other devices have a built-in rectifier in the power supply unit to enable them to convert AC to DC power.

Not all electronic devices use DC power; there are those that use AC power sourced directly from the power grid.

Resistance/Impedance

Resistance measures the ratio of voltage across an object to the current flowing through it. In fact, it measures the opposition of the current as it flows through the circuit.

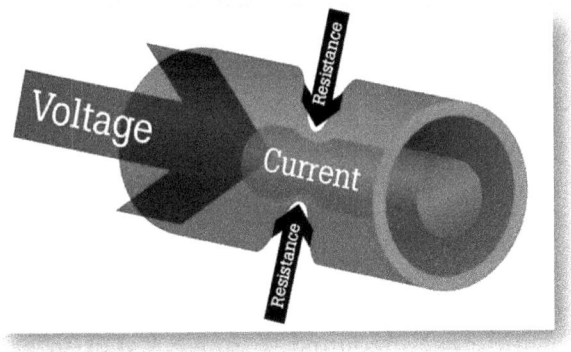

Resistance is measured in ohms, as denoted by the omega symbol (Ω). When current passes through a material, it experiences some resistance to a certain degree.

Conductor materials offer less resistance, and electrons can move freely through the material. For example, aluminum, copper, gold, and silver materials offer less resistance, while insulator materials have high resistance and restrict the flow of electrons through the material. Examples of insulator materials include glass, paper, plastic, rubber and wood. The higher the resistance is, the lower the current would be.

If the resistance of the circuit is constant, then you can use Ohm's law to determine the behavior of the material.

Ohm's Law

Ohm's law is one of the most practical laws of electrical circuits. It states that the current flowing through a conductor is directly proportional to the voltage across it provided temperature and other physical conditions remain constant. It measures the relationship between the potential difference and current flow as follows:

$V = I * R$, $I = V / R$, or $R = V / I$

Where,

- V is the voltage across the conductor (measured in volts) and measures the potential difference required to move a unit of charge between two points,

- I is the current passing through the conductor (measured in amperes), and

- R is the resistance of the conductor (measured in ohms).

For example, if a battery has a voltage of 10V with a resistance of 100Ω, then the current that is flowing through the circuit is 0.1A.

$I = V / R$, so:

10 V / 100Ω = 0.1 A

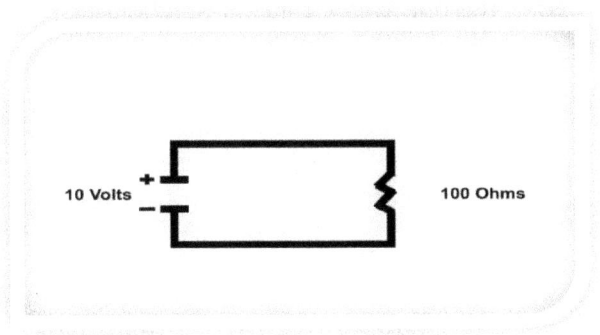

Resistance of the conductor (R) describes the ability of the circuit to resist or impede the flow of electrons moving through the circuit. This opposition results in the production of heat. For example, the current passing through the circuit in a tungsten light bulb heats up the filament, which acts as a resistance, thereby causing the bulb to emit light. Ohm's law holds true only if the temperature and other conditions remain constant. In some components, if the current increases, the temperature rises. In such instances, you can't apply Ohm's law.

Water Analogy for Ohm's Law

Since electrons are invisible, the water analogy can help to easily understand the flow of current within a circuit. The water that flows through pipes is similar to how current flows through circuit.

The water pressure represents voltage here, while the amount of water flowing through the pipe represents the current that flows through the circuit. If you have a bigger pipe, it offers less resistance as more water flows through the pipe (current) while pressure (voltage) is unchanged.

Similarly, if your water pipe is smaller, less water flows out of it with the same pressure.

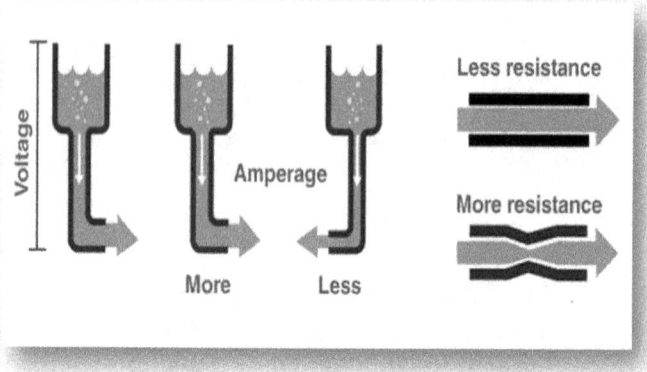

Watt's Law

Watt's law states that the power dissipated in a circuit is a product of its voltage and the current flowing through the circuit. It describes the relationship between the power, current, and voltage drop in a circuit.

Power = Voltage * Current

Or, $P = I * V$

For instance, if you have several 500-watt electrical devices, you probably want to know how many of them you can plug into your circuit without blowing the fuse. To determine how devices you can plug in, you need to first determine the total amount of current you can draw from the circuit. If the circuit has a 15A circuit breaker, and if the voltage required is 110 V, then the power dissipated in the circuit will be:

110 V * 15 A = 1650 W.

This is the available power for your circuit, so whatever device you plug into your circuit should be less than this. In our case, we can plug a maximum of 1650 W / 500 W = 3.3 (rounded down to 3) 500-watt electrical devices into the electrical circuit.

Variation of Watt's Law

Power is the rate at which a circuit uses energy and is a product between voltage and current: P = V * I

When electrons flow through a resistance, they collide with each other and with the atoms in the circuit. This collision results in the generation of heat, which, in turn, leads to energy loss: V = I * R

This results in a variation of Watt's law, and it is expressed as:

P = I * R * I, or: P = I^2 * R

Watt's law and Ohm's law use the same quantities, and you can combine both equations to come up with individual quantities. For example, if the power and voltage are indicated, you can determine the current flowing through the circuit using the Watt's formula.

Similarly, you can combine both Watt's and Ohm's law equations to determine resistance across an electrical circuit.

For example, if you have a 60-watt light bulb in a 120-volt circuit, you can calculate that the current flowing through will be 0.5 amperes.

Current = Power / Voltage = 60 / 120 = 0.5 amperes.

Common DC Circuit Terms

To better understand the next electrical laws, you need to know a couple of circuit terminologies such as nodes, loops, and branches:

> ***Circuit***: a closed-loop conductor that provides a path for the current to flow through. The diagram below illustrates a closed circuit.

> ***Node***: acts as a junction or connection terminal in a circuit that allows you to connect two or more elements together. It resembles a dot and forms a connection point between branches. For example, the above circuit has four nodes.

> ***Branch***: Consists of a single component or a group of components (like resistors, capacitors, etc.) which are connected between two nodes.

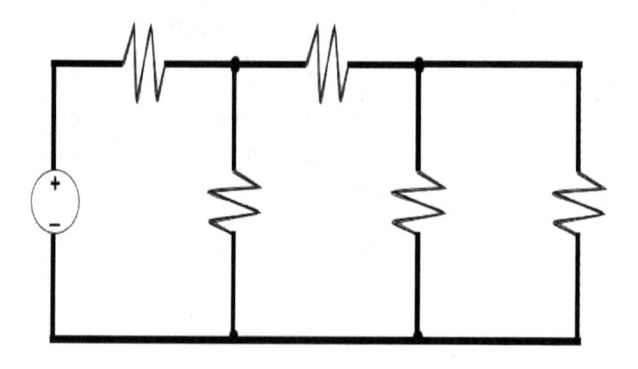

The following diagram has five branches (resistors plus battery).

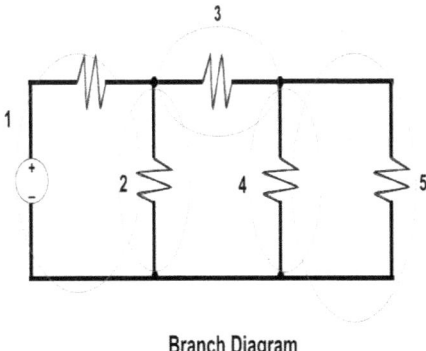

Branch Diagram

- ***Loop***: A loop is formed when a closed path goes through a node or any circuit element. A loop is, in fact, a pathway in which the current can flow. To draw a loop, select the first node as your starting point and draw a path through the elements or nodes until you go back to the starting point. Rule: you can go through each node/element only one time to form a complete loop. In the diagram below, there are six pathways, or loops, for the electrical current to flow.

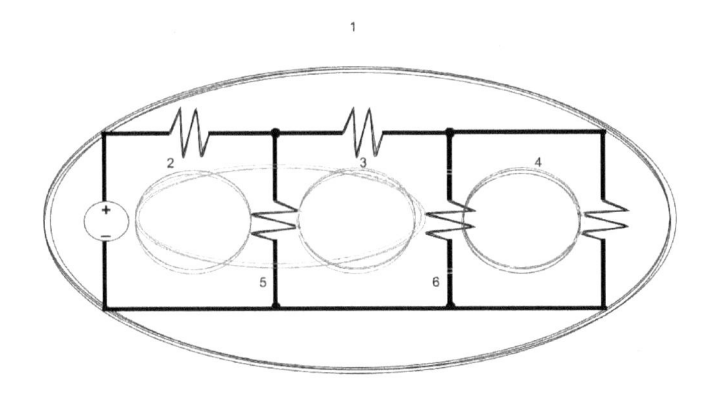

> ***Terminal:*** a point at which the conductor from a component of a circuit comes to an end.

Kirchhoff's Law

The concept of Kirchhoff's law is ideal for dealing with complex circuits. This law describes a relationship between the current flowing in a node and the voltage across a loop.

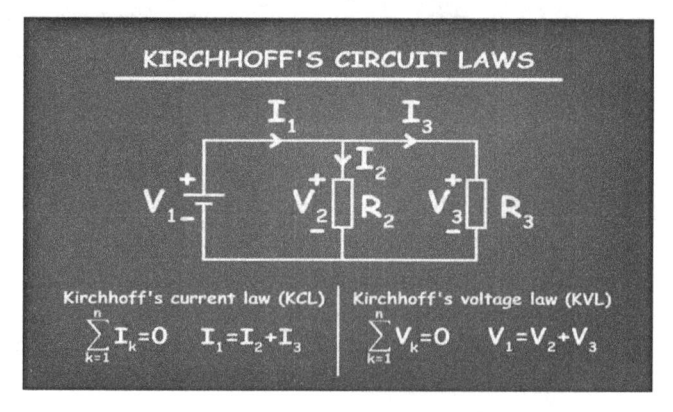

![Kirchhoff's Circuit Laws diagram showing KCL: $\sum_{k=1}^{n} I_k = 0$, $I_1 = I_2 + I_3$ and KVL: $\sum_{k=1}^{n} V_k = 0$, $V_1 = V_2 + V_3$]

Kirchhoff's Current Law (KCL)

Since all electrical components are connected together by nodes, the total current going through the node should be equal to the total current leaving that node. In other words, the *current in* is equal to the *current out*.

$$\sum I_{in} = \sum I_{out}$$

The sum of current entering into and out of the node must be equal to zero: $I_{in} + I_{out} = 0$

In this case, the node acts as a connector, or junction, between two or more elements or current paths. A *closed-circuit* path must exist for the current to flow in and out of the node.

Kirchhoff's Current Law

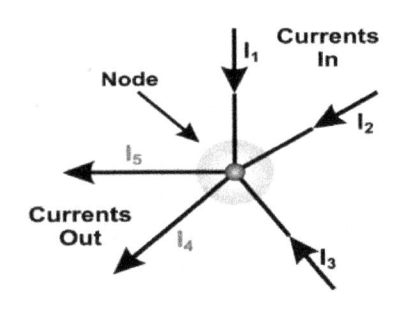

The current coming to a node is measured positive, while the one exiting is measured negative; thus, the sum of them is zero.

$I_1 + I_2 + I_3 + (-I_4 + -I_5) = 0$

Kirchhoff's Voltage Law (KVL)

This law states that the algebraic sum of all the potential differences (voltages) of all the elements connected in series in a loop is equal to zero. It also ensures there is the same voltage (potential difference) across each branch in the circuit. In other words, all the electrical components in the circuit have the same current flowing through them.

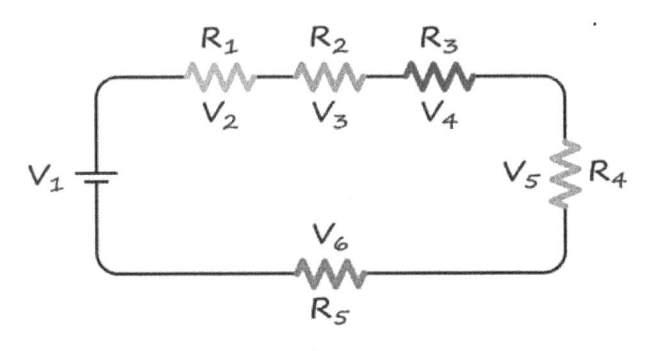

$V_1 + V_2 + V_3 + V_4 + V_5 + V_6 = 0$

As you can observe in the previous diagram,

$V1 = IR_1 + IR_2 + IR_3 + IR_4 + IR_5$

$V_1 + (-IR_1) + (-IR_2) + (-IR_3) + (-IR_4) + (-IR_5) = 0$

Or

$\sum V_{total} = 0$

As illustrated above, the voltage of the source (V1) is considered positive, while in the resistors (R1, R2, R3, R4 and R5), it is dropping (negative), and that is why the law states that the algebraic sum of voltage drops must be equal to zero.

Electrical Circuits

An electrical circuit consists of components that allow transmission, storage, and conversion of energy. The energy transmitted through the circuit flows through one or more sources and exits via one or more outlets. Every circuit has three basic elements: *voltage source, load*, and *conductive path*.

A voltage source consists of a battery that enables the current to flow through the circuit, while the conductive path allows current to flow through the conductive material. The load is an electrical device that consumes the power in the circuit. The conductive circuit establishes a relationship between the voltage source and the load. Circuits have a switch that you can turn on or off, and a fuse connected between the source and the load. Therefore, before you build your own solar system, you have to know about the circuits and how to successfully create one.

Components of a Simple Electrical Circuit

Different electrical components are connected together to form a circuit. Understanding these components will make it easy for you to build your own system.

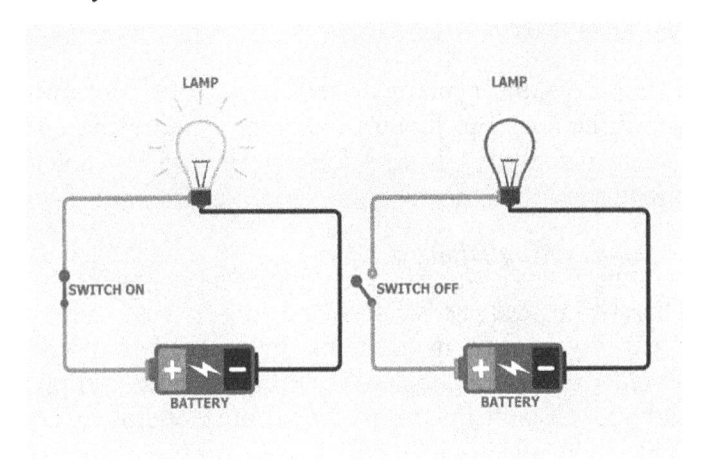

Switch

A switch can be in the form of a pushbutton, momentary, or rocker, and it interrupts electrical current when you turn the circuit off.

Resistor

A resistor is a basic component that controls the voltage and how current flows through the circuit. Before designing your circuit, it is important to know the size of the resistor.

You can use Ohm's law to determine the amount of resistance in a circuit if you know the voltage and the amount of the current (amperage) passing through the circuit.

Capacitor (voltage force)

A capacitor is another basic electronic component that stores electricity and is also responsible for discharging electricity back into the circuit when the voltage drops. It acts like a rechargeable battery.

Diode

This allows the generated electricity to flow in only one direction. The diode's main role is to route and block electricity from flowing in the opposite direction or in an unwanted path in the circuit.

Light-Emitting Diode (LED)

The LED acts as a standard diode that allows electricity to flow in only one direction. The insides of the LED component have both an anode (+) and cathode (-), and by the electrical current flowing, it emits light. Electricity always flows from the positive side (anode) to the negative side, and not in the opposite direction.

Transistor

A transistor acts as a tiny switch, and, when triggered by an electrical signal, it turns the current on or off. It can also amplify signals.

Types of Circuits

If you have two or more electrical components to connect in a circuit, there are certain situations in which the current and voltage of the circuit are critical to know. Electrical components, such as resistors, form a two-terminal object (a circuit with two endpoints).

1. Closed Circuit

A closed circuit works like a circle that allows electricity to flow from one end of the circle all the way back to the starting point, forming a complete loop. The current flows from the positive terminal to the negative terminal of the capacitor uninterrupted. As illustrated below, a closed circuit provides a low resistance path for the electrical current to pass through.

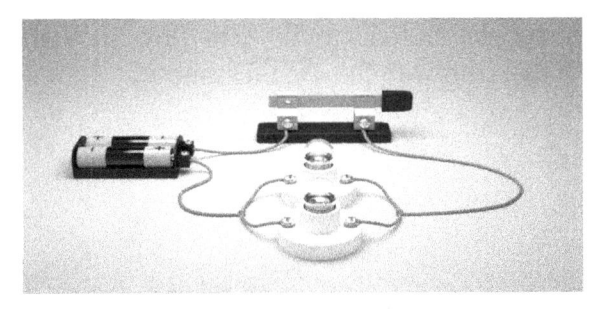

2. Open Circuit

When you switch off the circuit or if there is a fault in your electrical wire, the current stops flowing, forming an open circuit. In such cases, there is no continuity of current flow because of the broken wire as depicted below.

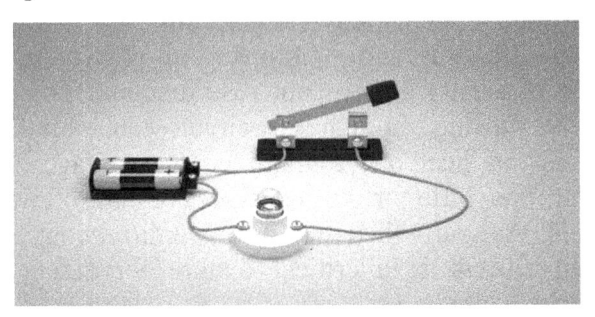

Turning the light switch on or off opens or closes the circuit that connects lights to the energy source. If you disconnect the battery, it creates an open circuit. Open circuit voltage is considered a crucial feature in order to size main PV equipment.

3. Short Circuit

When you directly connect two points that are not supposed to be connected, they form a short circuit. For example, if you connect two end terminals of a power supply, the electricity will flow through the conductive path with the least resistance.

SHORT CIRCUIT

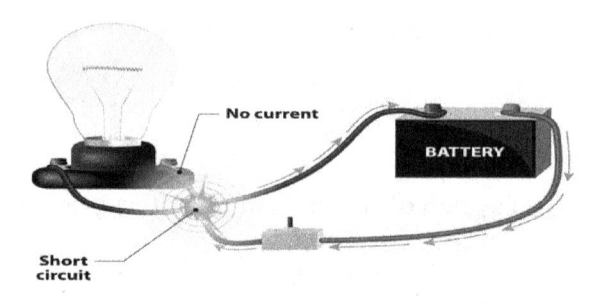

The current can bypass the parallel conductive paths and flow through the direct connection path; however, in real life, this can shut out your power supply, which is usually due to your fuse function. The high current flowing through the circuit can emit large amounts of heat due to high energy voltage. This circuit has little or no resistance to the flow of current. As you will notice in this entire book, short circuit current is a critical feature for sizing PV equipment and wires.

Types of Connection (Series, Parallel)

Elements of a circuit may be connected in series, parallel, or both.

Series Connection

In a series connection, the electrical components follow a single electrical path, and the same current flows through each of the connected components. The voltage across the circuit is equivalent to the sum of the voltages across each of the connected components (as mentioned in Kirchhoff's voltage law).

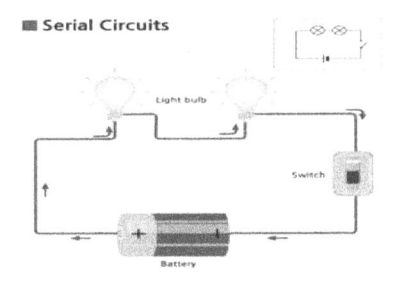

Components of a series circuit are connected in line with the power source, and the current is constant throughout the circuit. If you open or break a series circuit, the entire circuit will stop operating. For example, if one of the light bulbs connected to a string of lights in a Christmas tree burns out, the entire series of the Christmas tree lights will not work until you replace the broken one.

A series circuit consists of several resistances that are connected one after the other forming an end-to-end connection.

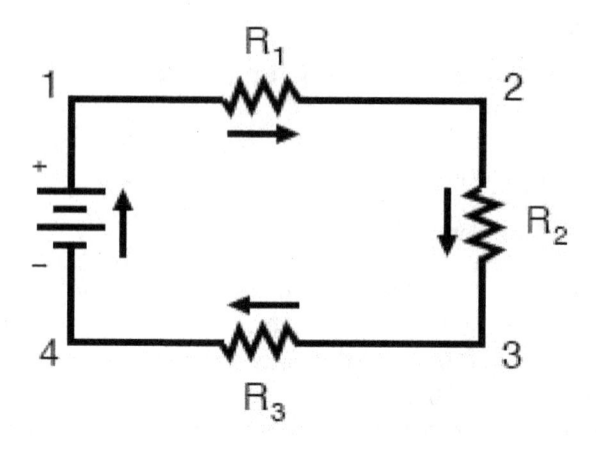

As depicted in the diagram above, the current flows in a clockwise direction from point 1 to 2, 3, 4, and back to point 1. The resistors (R_1, R_2, and R_3) are connected in a single chain to the battery:

$R_{total} = R_1 + R_2 + R_3 + \ldots R_n$

The voltage supplied in the circuit is the sum of individual voltage drops across the resistors, while the current passing through all elements is equal.

$V_{total} = V_1 + V_2 + V_3 + \ldots V_n$

$I_{total} = V_{total} / R_{total}$

Parallel Connection

In a parallel connection, electrical components connect along multiple paths, and the voltage across each of the components is the same. Components connected in parallel have a constant voltage and branch off from the battery.

Parallel Circuits

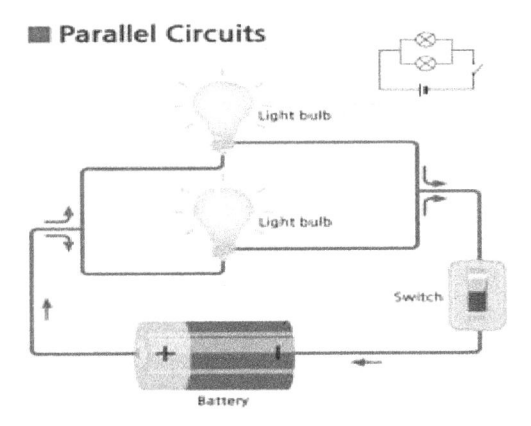

The current that flows through the circuit is equal to the sum of currents across each of the components (resistors).

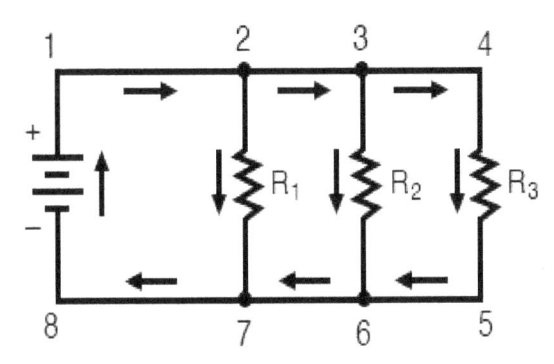

Again, the circuit depicted in the previous page has three resistors, but the current has to flow through multiple paths. The first path flows from point 1 to 2, 7, 8, and back to point 1. The second path flows from point 1, 2, 3, 6, 7, 8, and back to 1. And the third path flows from 1, 2, 3, 4, 5, 6, 7, 8, and back to 1. Each path flowing through R_1, R_2, and R_3 is called a *"loop."*

$1/R_{total} = 1/R_1 + 1/R_2 + 1/R_3$

Series-Parallel Configuration

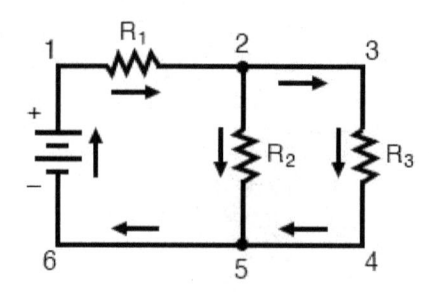

The above series-parallel configuration consists of two loops to allow the flow of current:

The first loop allows current to flow from point 1, 2, 5, 6, and back to point 1. And another flows from 1, 2, 3, 4, 5, 6, and back to 1. Both currents flow through R_1 (from point 1 to 2), while R_2 and R_3 are configured in parallel to each other; therefore, while the total voltage of this circuit is equal to the sum of voltage in R_1 and either voltage of R_2 or R_3, the current passing through R_1 is equal to the sum of the currents of R_2 and R_3.

Electrical Load

Any device that consumes electrical power is known as an electrical load. Devices such as light bulbs, laptops, TVs, refrigerators, cell phones, etc., are part of an electrical circuit and will consume electrical energy. An electrical load transforms the electricity into other forms of energy, such as heat, light, or motion.

The term "load" also describes the power requirements of each supply unit or the amount of current flowing through the circuit. The type of load consumed depends on the demand, diversity, power, and utilization of your system. It can either be Resistive, Inductive or Capacitive.

Resistive Load

A resistive load consists of lights and heating elements like ovens, incandescent lights and toasters. They consume active power which only flows from the voltage source to the load, not in the opposite direction.

In addition, the voltage and current waveforms are usually in phase with each other and reach the peak at the same time. As depicted below, sinusoidal waveforms are characteristic of AC power and are due to the oscillating movement of stationary coils in electrical stations:

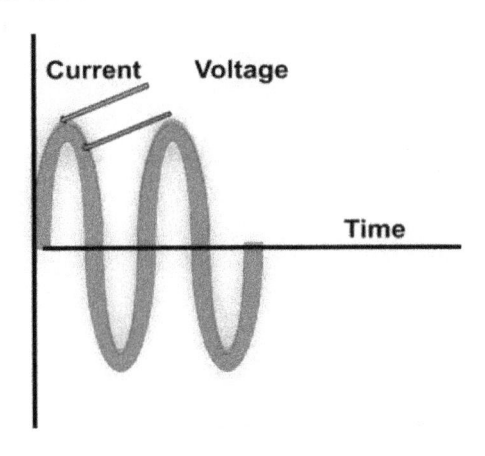

Inductive Load

This type of load only consumes reactive power. Examples include: electrical motors, transformers, and generators. The coil in the load stores magnetic energy as current flows through it. A number of household items with moving parts, such as washing machines, vacuum cleaners, dishwashers, air conditioners, and compressors, use this type of load. Some of these devices need peak surge current to start, which must be taken into account while calculating your daily electricity consumption.

The current waveform lags behind the voltage wave causing lagging of the power factor of the inductive load. Both voltage and current are out of phase and lag by 90 degrees. In such cases, when the current is at zero, the voltage is at its maximum.

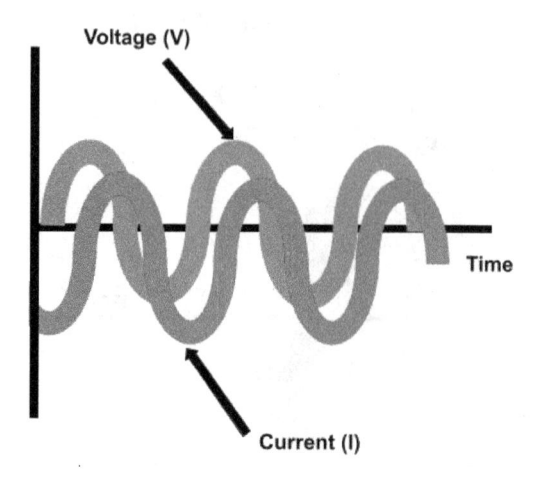

Since inductive loads consume reactive power, the power can flow either from the load to the source or from the source to the load.

Capacitive Load

In this type of load, the current wave reaches its maximum peak point before the voltage wave. The two waves are out of phase and the current waves are always leading by 90 degrees. Capacitive load is widely used in capacitor banks and in three-phase induction motors.

Since the capacitive load doesn't exist in the form of a stand-alone format, there is no specific device that is categorized as capacitive. Large circuits use capacitors to control power usage. Electrical substations use capacitors to improve the power factor of the system.

Chapter Summary

Having knowledge of basic electrical rules, circuits, and formulas is essential if you want to build your own solar power system. Electricity rules will enable you to know how to connect different electrical components based on the amount of power you want to draw. There are also different types of circuits, and they have different ways of functioning.

A proper understanding of how electricity is used for different household purposes will help you throughout different steps in your solar journey. In the next chapter, you will get familiar with essential tools for solar power design, installation, and maintenance.

DIY SOLAR POWER FOR BEGINNERS

CHAPTER THREE: Essential Tools and Equipment

The shift to solar power has led to the invention of highly specialized tools for solar power installation. Though some of these tools are easy to find around your home, there are others designed specifically for solar power installations. These tools range from simple devices that allow you to disconnect the solar connections to highly complex tools.

If you want to install the PV system yourself, then you need to learn about the basic tools required to set up your system. DIY installations lower your solar power costs, but if you don't have the basic installation tools and want to hire a contractor, your upfront costs will rise.

In addition to having the necessary installation skills, you also need to consider your safety. As with any DIY situation, accidents and injuries can happen, and you obviously want to avoid that; therefore, before you start your DIY installation project, you must collect all the required tools. In some cases, you have to hire the services of a professional electrician to help you with more complicated procedures, such as wiring and bending of the conduit. But if you have the basic electrical skills, you can utilize them. Remember to observe all the safety precautions!

Let's have a look at some of the common tools needed for solar installation.

Safety Tools

The installation of solar panels requires heavy lifting, climbing on the roof, and electrical wiring; thus, your safety is of utmost importance. Some of the protective equipment you need to consider includes the following.

Gloves and Boots

Always wear gloves and closed-toe boots to avoid cuts, scrapes, and other injuries when installing panels. Boots with flat soles provide you with the grip required when working with shingle roofs, tiles, or metals. You can also wear long-sleeved clothes to prevent the burning of your skin when working outside or on the roof.

Goggles

Wear eye protection glasses and other protective gear to prevent eye injuries when working. Since you will be exposed to direct sunlight while working on the roof, some sunscreen lotion is a good idea as well.

Roof Anchors

Roof anchors protect you from falling when installing a roof-mount system. Always attach a safety harness to the anchor when working on the roof to provide yourself with an extra layer of protection.

Scaffolding

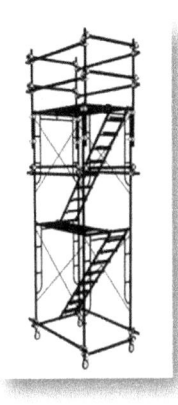

A scaffold helps you climb to the roof and can act as a boom-lift for getting the equipment to high places. You can make a temporary scaffold structure that aids in your installation of panels at higher places. Regular ladders can be also used in simpler projects.

Guardrails

When installing solar panels on a roof, especially in more extensive projects, you are at risk of falls. Thus, there's a need for having solar safety hacks like temporary guardrails. A roof safety harness tool also protects you from falling.

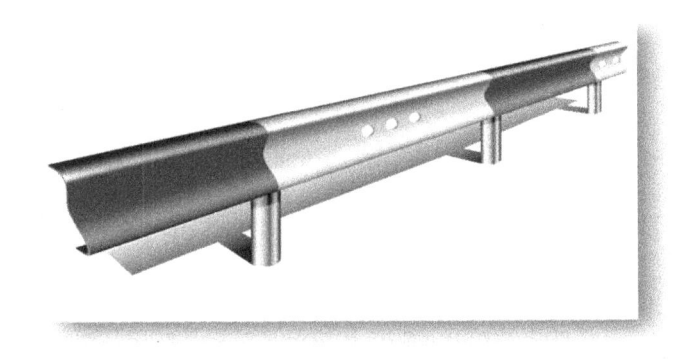

Ladder

You need a sturdy ladder to help you climb up and down with ease. The ladder should extend at least three feet above the edges of your roof.

Multimeter

Use a multimeter to check whether there is voltage before you work on the system. Ensure the conductors and terminators have zero voltage before starting any installation—that way, you'll avoid an electric shock.

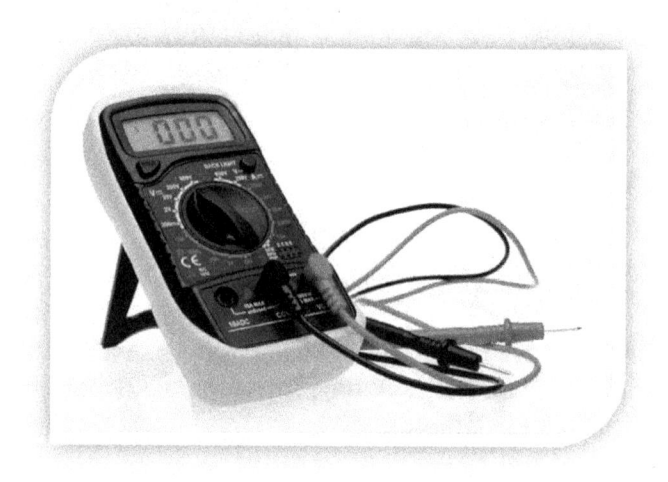

Power Tools

Pitched and sloped rooftops require different mounting systems and, consequently, different tools. Additionally you will need a majority of these tools for other solar components installation. Some of the common solar installation tools you should consider, are listed and explained as follows:

Cordless Drill

A cordless drill is a powerful tool that simplifies your DIY work. It helps you drill pilot holes, drive in lag screws, and tightly fasten them into the roof. A cordless drill has a variety of functions, and every homeowner should have one. A general-purpose drill with a 12V or 18V battery is suitable for most solar purposes.

Impact Driver

Just like the cordless drill, impact drivers are efficient in driving long deck screws, tightening module clamps, and fastening racking bolts on surfaces. An impact driver is built for driving screws, but cannot drill. It is suitable when dealing with large screws and bolts because it exerts an extra rotational force on hard material.

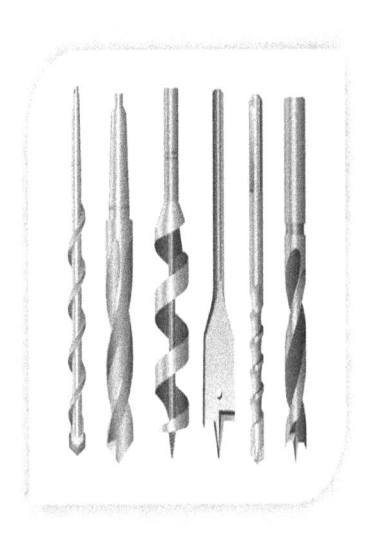

Drill Bits and Sockets

Drill bits act as cutting tools that create holes of different sizes and shapes on different types of materials. Always choose a drill bit larger than the hole size you want to create.

The twist bits are the common drilling bits for plastic, timber, and metal materials, and you can drill them with your hand or use an electric drill.

If you want to drill into concrete, stone, or brick material, you should use masonry bits.

A drill bit socket is a tool that adapts sockets for use in a drill. They have an adapter fitted on a tapered shank drill to a tapered hole that is larger than the created hole size. The sockets fit on the nuts or bolts to tighten the drill bits as depicted below:

Caulking Gun

This is a tube filled with cartridge material (roof sealant) that seals up any gaps or holes left after fastening screws on your rooftop.

The cartridge material used may be either silicon or latex, and you can use it to bond together a range of materials such as glass, metal, or ceramic.

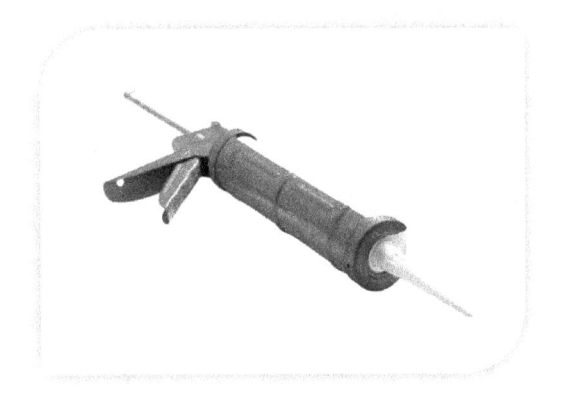

A caulking gun regulates how much caulk gets out of the tube when you squeeze it. The roof sealant prevents any leaks due to drilling installation holes. When buying a roof sealant, go for one that is appropriate for your roof sealing purposes.

Jigsaw

This is a powerful tool that allows you to cut rails after installing the solar modules, for instance.

It enables you to have more control when cutting complicated patterns or shapes to avoid damaging your roof.

Reciprocating Saw

A reciprocating saw is a handheld tool that allows you to quickly cut through a number of materials.

The saw has a large blade just like a jigsaw blade, and an oriented handle that allows you to comfortably cut materials on vertical surfaces.

A chargeable saw is preferred on the roof.

Hole Saw/Hole Cutter

A hole saw is a ring-shaped blade that makes holes on a surface without cutting the core material. It is suitable for drilling, and the hole creator has a pilot drill bit at the center to prevent the saw teeth from moving. You can use this tool to cut through thin metal plates and roof material, etc.

Screwdriver

A screwdriver allows you to tighten or loosen different types of screws. Although impact drivers have dominated most screw driving applications, manual drivers might come handier in certain situations.

Pliers

Crimping pliers have jaws that allow you to grip objects when you squeeze the two handles together.

Though most pliers are designed to perform general-purpose work, there are those designed specifically for certain purposes.

If you have long nose pliers, it can bend wires or squeeze out tight spaces. Pliers with sharp edges can shear through thick electrical wires. You can also use pliers to grip objects when doing the installation. Holding wires with pliers helps prevent electrocution.

Measuring Tape/String Line/Chalk Line

A measuring tape is obviously needed for your project. For instance, you can measure the distance between the drilled holes on your panels and mark corresponding areas on your rooftop to know where to drill the holes. A chalk line tool marks straight lines on a flat surface. You can use it to lay straight lines between two points and ensure the panels are mounted on the chalk lines. A string line ensures that your solar panels are installed in perfectly leveled squares. Everyone appreciates well-organized solar panels on the roof.

Roof Sealant

A roof sealant helps prevent leaks from the drilled holes. Always make sure to buy a sealant suitable for your roof material.

Wiring Tools and Equipment

You will need to know how to use several tools and equipment specifically used for wiring.

Junction Boxes and splices

This is an electrical box that encloses electrical wires and cables. If you cut electrical cables or splice the wires together, then you need a junction box to protect against short-circuiting of the wires.

These safety measures prevent electrical shock from live wires in your home. All wires connected to your switches and lights should be enclosed in a junction box which will serve as an enclosure for all spliced wires. Each box consists of a connection splice that accommodates two or more circuit cables. Cables entering the box are secured by either conduit connectors or cable clamps.

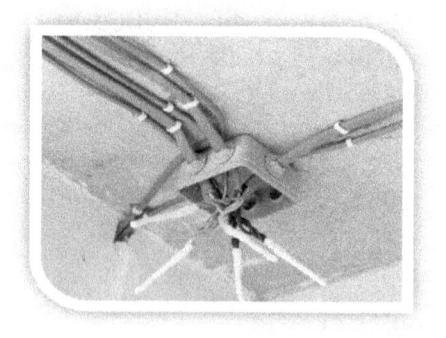

When buying junction boxes and splices, ensure they meet the voltage and current requirements of your circuit. Always use approved splicing devices or insulated lugs.

AC Breakers

An AC circuit breaker prevents damage to your appliances by cutting off the power supply when it detects an overload. It functions by interrupting current flow when it detects a fault. The circuit breakers can also import and export power from your equipment. Below you can see a sub panel containing a few circuit breakers:

Subpanels

This is an essential component that helps you add additional circuits in your home when slots of the main panel circuit breaker are full. It extends the distribution of power to specific areas in your home. A subpanel has its own breakers, which makes it easy to extend your wiring using multiple branch circuits to your home or any other building if they're far away from the main panel.

PV Meters

If you're installing a grid-tied or a hybrid solar system, you have to install a dedicated PV meter that measures the amount of energy generated by your solar array. In most cases, you have to install the base of the meter socket, and once your solar system is approved, the utility company will install the meter face and activate your PV system.

Conduits

A conduit is a tube used to protect your electric wiring from moisture, impact, and chemicals in exposed areas. The plastic sheath enclosing the wires is prone to damage. To avoid this, you can pull single strands of wires into a metal conduit to protect sensitive electrical circuits. Electrical metallic tubing (EMT) conduit is rigid steel raceway which not only protects the wires but also grounds the panels.

Always make sure you have the right size of conduit before starting the installation process. Once you place the conduit correctly, you can tighten the joints using a set of channel locks.

Channel Locks

Channel locks look like pliers that you can use to grab, hold, and turn nuts or bolts.

The tool is also great for crimping metal objects or the end of pipes. They come in a set of 6.5, 9.5, and 12-inch pliers.

Equipment Grounding Conductor (EGC)

This is a bare copper wire used to ground the solar components. It connects solar panels, EMT conduits, and other metal enclosures together.

EGC provides a path for connecting electrical components to the ground to avoid ground faults.

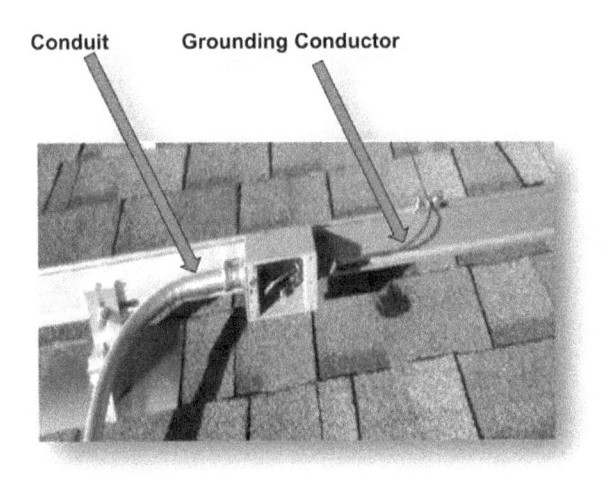

Wire Cutters

A wire cutter allows you to cut wires made of aluminum, brass, copper, iron, and steel material. The insulated handles help prevent electrocution from the wires while working, and they provide a comfortable grip. A diagonal flush cutter enables you to cut wires at an angle close to the base.

Wire Strippers

Wire stripper is another must-have tool, as they allow you to remove insulation from an electrical wire when you want to make a contact. There are different types of wire strippers with varying notch sizes.

Wire Crimpers

A crimping tool fixes connectors at the end of the cable. Though it looks like a pair of pliers, it joins together two pieces of a metal plate. There are three types of crimping tools: ratcheting, hydraulic, and hammer.

A ratcheting crimping tool allows you to secure insulated wire connectors, terminals, and heat shrink butt splices. You have to apply a threshold pressure to join the connectors on the two pieces of the wire. They come in a variety of sizes, and you can interchange them to crimp connectors of different widths. Before buying a crimping tool, confirm the type of wire or cable it can crimp.

A hydraulic cable crimping tool is a special crimping tool that crimps terminals of wire ropes and conjoins two pieces of metal together.

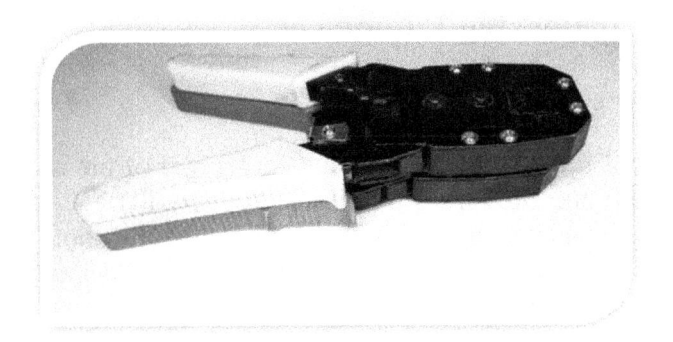

A hammer crimping tool can crimp connectors, terminals, and splicers together. A hammer is needed to apply force to the crimping tool.

Fish Tape

This is a useful tool that allows electricians to route a wire through the wall or pull it through an electrical conduit. The fish tape has a flat, long, and thin steel wire that is wound up inside a round-shaped wheel. The fish part allows you to attach the wire, then pull the wire through the conduit.

A conduit is a plastic or metallic pipe similar to a plumbing pipe that protects your electrical wiring.

Torque Wrench

A torque wrench is a unique T-shaped tool that allows you to use a specific torque to fasten bolts, nuts, and lag screws on solar panel rails. It is ideal in situations where the tightness of screws and bolts is very important. A high-quality torque wrench helps you tighten bolts and nuts on the rails. Though this may consume a lot of time, it ensures the components of your setup are in place and tightly fixed.

Solar Racking Equipment

Roof Rafters

A roof rafter is a structure that forms part of the roof design and runs from the hip of your roof (ridge) to the wall plate. In fact, it is not really a solar tool or piece of equipment; however, it provides base support for mounting solar panels.

Your decision on where and how many panels to install is significantly influenced by the position of the rafters on your roof.

Rails

Rails are mounted on the roof to support solar panel rows.

For installing the solar panels you need to place each of the panels vertically or in portrait position, and then use two rails with clamps to secure the panels on the roof or on the ground. The rails are secured to your roof using screws or bolts.

Roof-Mount Flashings

Roof flashing is a thin metal material in the form of galvanized steel, aluminum, or copper. Its main function is to avoid any water leakage around the holes drilled in the roof in order for the rails to be secured.

The following schematic depicts a racking system and the relationship between flashings, rails, clamps, and solar panels:

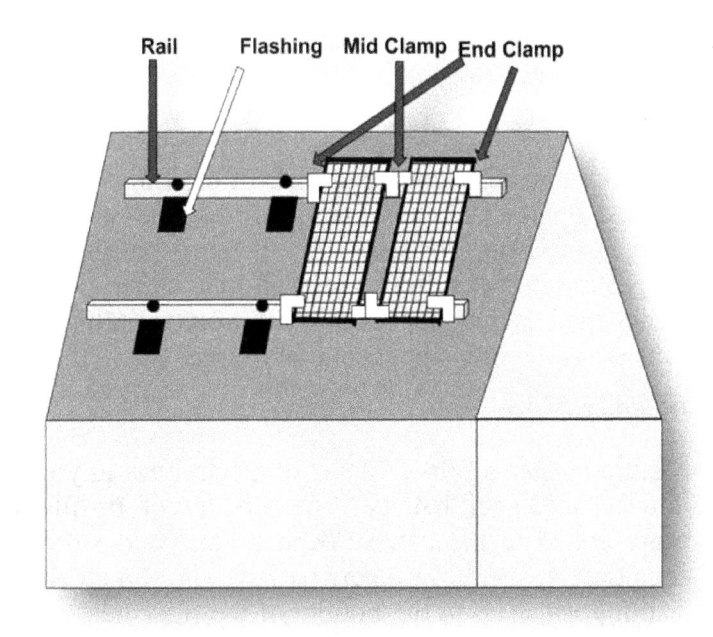

End Clamp, Mid Clamp

Clamps allow you to hold or position the panels on the rail.

While end clamps secure the panels at rail ends, mild clamps are located between two panels and keep them attached to the rails.

Solmetric Sun Eye/Pathfinder

These are solar assessment or shade analysis tools that provide accurate measurements of the amount of solar energy generated per day, month, or on an annual basis.

These tools can also measure shading patterns within a particular area.

Battery & Maintenance Tools

Hydrometer

This measures the relative density of electrolytes in a flooded lead-acid battery to determine the state of charge of your battery. If there is a higher concentration of sulfuric acid in the battery, that means there is a higher level of electrolytes. A higher density results in a higher state of charge.

Distilled Water

Distilled water refills the level of electrolytes in your flooded lead-acid battery. The amount of distilled water added to the battery depends on your battery condition. If you have a new battery, you should add distilled water up to the bottom of the filler tube. An old battery requires distilled water up to the level of the electrode. You can use a small flashlight to view the electrolyte level in the battery.

Baking Soda

If sulfuric acid from the battery leaks on the surface, you can pour baking soda on the spot to neutralize acid spills.

Funnel

This is used to guide liquids through the small opening of the lead-acid battery case for refilling distilled water into the battery.

Rubber Apron

Before handling any dangerous and toxic chemicals you have to wear a rubber apron to protect your body and clothes against any spills. Even a drop of sulphuric acid or any other powerful chemical could cause serious injury to your skin.

Rubber Gloves

If you work with toxic chemicals and other harmful detergents, you have to wear gloves to protect your hands.

Chapter Summary

Effective design, installation, and maintenance of solar power systems requires a deep knowledge of solar equipment; therefore, the most essential solar-related tools were explained in this chapter to make your solar plan more feasible.

If you are interested in various features of different types of solar power systems, move on to the next chapter.

DIY SOLAR POWER FOR BEGINNERS

CHAPTER FOUR: Types of PV Systems and Components

The ever-increasing technological advancements in the solar field have led to the introduction of several distinct systems harnessing solar power to be utilized for residential and commercial purposes. This chapter will help you understand different PV systems' indications and recognize their differences in order to select the one that best suits your situation.

When you install solar panels, the generated energy must be converted to usable electricity. This can be accomplished through various ways: connecting your system to the grid, going off-grid completely, or having a hybrid system that combines the two systems.

Grid-Tied Solar Power System

A grid-tied (on-grid) solar power setup is a system that connects to an electrical power grid. It first satisfies all your power needs before delivering the excess power to the electrical grid. When the solar system cannot satisfy your power demands, especially at nights or on days without sunlight, you can draw power from the grid. It is the least expensive system of generating energy since no battery backup is required.

The excess energy produced is directly sent to the power grid utility company. In this case, the electrical grid acts as the battery backup for your system.

The amount of power fed to the electrical grid slows the rate at which the electric meter measures the power usage. It causes the meter to spin backward. In other words, if the meter spins backward, it is an indication that solar power is being fed into the grid.

When unused power is delivered to the grid, the utility company uses "net metering" to credit homeowners with the per kWh price they pay for energy consumption when drawing power from the grid.

A grid-tied system is the preferred solar power system on the market because it is the least expensive and least complicated PV setup. Homeowners can enjoy more than fifteen years of profits after the break-even on solar system investment.

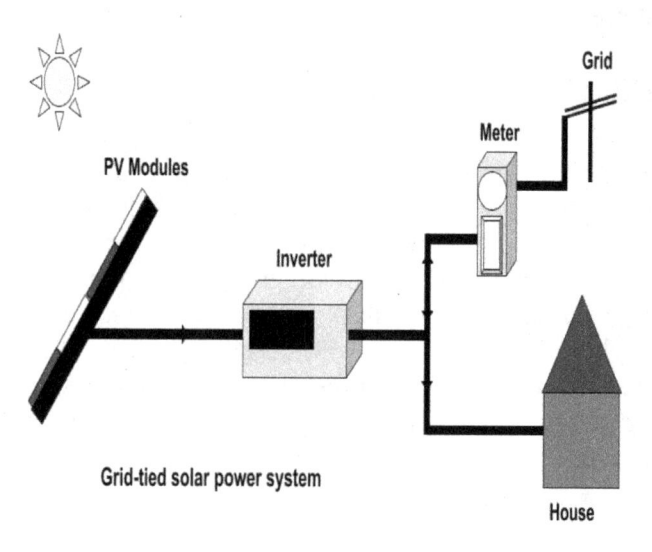

Grid-tied solar power system

The panels should be installed in a place where there is maximum sunlight exposure. This can either be a roof or ground mount. Proper wiring is required to transport power between the connected solar equipment, your home, and the local electrical grid.

Advantages of Connecting to the Grid

- **Lower initial cost:** grid-tied systems don't need any special equipment when compared to an off-grid system. Since the system uses less equipment, there will be less installation work, and this reduces your labor cost. This makes it the least expensive type of PV installation.

- **Net metering:** you take advantage of net-metering to save on your electricity bills and limit wastage of energy.

- **Backup energy source:** it helps improve electrical grid efficiency during its peak usage period.

- **Reliable:** grid-tied systems are more reliable and require little maintenance when compared to other types of PV systems. The fewer types of equipment required mean there is a lower potential for failure.

- **Short payback period:** installing a solar power system is a decent investment for your home. Grid-tied systems have a short payback period when compared to battery-backed systems. On average, you can break even the total installation cost within three to five years.

- **Passive income:** when you sell excess power to the grid, the utility company pays for the energy received using the current per kWh price.

Disadvantages of Grid Connection

- **Not suitable in undeveloped countries:** installing a grid-tied solar system is not feasible in undeveloped countries or remote areas without power lines.

- **No power in case of grid outages:** if the main power grid goes off, the solar grid-tied system will shut down—therefore, no energy is delivered to the system. And since there are no connected batteries to the system, there will be no backup power during a power outage.

Components of a Grid-Tied System

Photovoltaic (PV) Solar Panels

A grid-tied PV system consists of an array of panels that are electrically connected or tied to the local grid. This setup is permanently connected to the electrical grid.

The number of panels required is determined by your decision on what share of your total electricity consumption your system is supposed to generate.

You also need racking or mounting equipment that ensures your solar panels are in place. Racking also ensures there is enough ventilation to allow cooling of the panels.

Solar Inverter

PV panels generate DC power while most home appliances and other electronic devices use AC power; therefore, an inverter converts the DC electricity into AC power.

A grid-tied inverter also analyzes the real-time flow of energy to determine whether the generated energy output is enough for home use or needs to be exported. Always look for a quality inverter that fits your budget and your electricity needs. Additionally, you have to consider the efficiency with which the inverter converts solar power to alternating current.

Electricity Meter

The electricity meter records electricity flowing into and out of the local grid. You can use twin kWh meters where one records the amount of electrical energy consumed, and the other records the amount of energy sent to the local grid.

A bidirectional kWh meter can also be used to record the net amount of power drawn from the grid.

Electricity Grid

You need the utility grid to connect your solar system; otherwise, without the grid itself, the system will not be a grid-tied PV system. Further, a grid-tied system is not an independent power supply source.

If there is an interruption of the main power supply from the grid, you might not have lights even if there is enough sunlight. This is because of your grid-tied inverter that shuts the power off for safety purposes.

Wiring/Cables

Wiring and cables transport power generated by panels to the inverter, and then to your home or to the electricity meter (net metering).

How Changes in Seasons Affect Grid-Tied Systems

Changes in the seasons and the time of the day affect the transfer of energy from homes to the electrical grid. When there is a change in the season, there is also a change in the amount of sunlight that hits the solar panels.

These changes determine whether your grid-tied system will import or export power to or from the electrical grid.

Grid-Tied Solar Power System with Battery Backup (Hybrid Solar System)

This is another grid-connected solar system that has battery backup within its design. This photovoltaic solar system works closely with a local electricity grid company to supply you with power if the battery runs out of power. You can also sell surplus electricity to the utility company.

The battery meets short-term demands during nights or days with bad weather without having to draw power from the grid at an extra charge. Adding batteries into your grid-tied solar system requires more components, which results in an increase in the installation cost while reducing the efficiency of the system.

The energy generated by panels passes through the hybrid inverter, which can decide whether to convert it to usable AC power and/or direct it to the grid or to store it in the battery backup. The hybrid inverter is a smart, two-in-one device: a solar inverter and a battery charger/inverter.

The main advantage of having a hybrid system is that you will always have electricity even at night, on cloudy days, or when there are blackouts. If you use up all the power in the battery and your panels cannot generate more power, you can draw power from the electricity grid to meet your power demands. There are two main configurations with which you can set up the major solar components in a hybrid PV system: AC-coupled and DC-coupled. Depending on which one you choose, both the wiring and the converting components might significantly differ from each other.

Components of a Hybrid System

Solar Panels

Depending on your energy requirements and the efficiency of your system, you can choose panels that help meet your specific demands. You have to determine the right size for your system to meet your energy needs. As with any PV system, a specific racking system is required for panels' installation.

Hybrid Inverter

As mentioned above, a hybrid inverter combines two separate devices in one unit:

A battery charger to charge the storage, and a solar inverter to convert the panels' generated power to AC electricity.

Due to having more complicated functions, hybrid inverters are more expensive and less efficient than grid-tied inverters. If you are anticipating adding batteries to your existing grid-tied system in the future, it is wise to consider a

hybrid inverter when planning your system. A hybrid inverter, in this case, works similarly to the traditional grid-tied inverter. Once your budget lets you add batteries, your hybrid inverter acts as a battery charger/inverter, as well as a solar inverter. Alternatively, you can add a battery charger/inverter separately to your grid-tied inverter in case you want to add a backup to the existing, totally grid-tied setup.

Battery Bank

Depending on the size of the batteries, they can store power for use ranging from several hours to several days. If you often experience interruptions of grid power supply, especially during bad weather conditions, batteries will be of great benefit.

Tesla Powerwall is one of the most popular home battery systems due to its capacity of 14kW, which can store half the average daily energy consumption for homes in the US.

It allows you to keep lights on 24/7. Additionally, Powerwall's unique technology allows you to use all the stored energy in the battery without damaging it. You can check it by scanning the code above:

Charge Controller

The basic function of all charge controllers is to prevent battery overcharging, as well as blocking reverse current from batteries. Some charge controllers on the market provide extra functions, such as controlling proper voltage in relation to temperature and battery type, as well as protecting the battery against over-discharging.

harge controllers are required in DC-coupled hybrid systems, where the solar panels' generated DC power is used to directly charge the batteries.

Wiring

The cables transport the power from the panels to the charger controller, then to the inverter, and then to your home, battery, and the net metering. Wire configuration depends on your solar array configuration (series and parallel) and whether the AC- or DC- coupling battery configuration is used.

In an AC-coupled wiring, the generated DC power from panels is converted to AC power for household use and/or sent to the grid.

Then another inverter/charger converts the AC power to DC again to charge the batteries. This same device converts the battery's DC power to usable AC power again.

In DC coupling, however, the DC-generated power passes through a charge controller to charge the battery, and then an inverter discharges the battery to generate AC power for household uses. Both types of battery coupling have their advantages and disadvantages and will be covered in detail in chapter eight.

Net Metering

This type of meter records the amount of power sent to and received from the electrical grid company for credit.

Advantages of Hybrid Systems

- **Continuous power supply:** hybrid systems ensure you can have access to power supply 24/7. The backup battery provides you with power for night use or during cloudy days. The electricity grid provides extra power that meets your energy needs.

- **Saves you money:** although the upfront cost of installing the system is considerably high, it saves you a lot of money in the long term. It lowers your electricity bill and makes the investment cost worth it.

- **Stores excess power:** the battery stores the excess power for use later, reducing your reliance on the utility company.

- **Easy-to-switch power load:** you can easily switch from using battery backup power to using grid power at any time.

Disadvantages of Hybrid Solar Systems

- **Higher upfront cost:** the installation cost of a hybrid system is higher than that of a grid-tied one due to the additional components. In addition, the price of a backup battery is significantly high.

For instance, the Tesla Powerwall 2 system including battery, inverter, and other component will roughly cost you 10,000 to 16,000 USD (before incentives), and this is excluding solar panel costs and installation expensive.

- **Requires a larger installation space:** due to the extra components needed for installation, you need a larger installation space for your system.

- **Not suitable for remote areas:** there are no grid power lines in remote areas, so you can't use hybrid systems in these places.

All-in-One Solar Power Systems

Recently, all-in-one PV technologies have been introduced to the market, like Growatt and Renogy. Also known as inverter chargers, with a built-in charge controller, these devices provide lower upfront and installation costs, and require less space.

The system is a combination of AC charger (to charge your batteries by grid or a generator), solar inverter (to convert PV's generated power to usable AC power) and a solar charge controller (to manage battery charging and discharging).

These space-efficient technologies are available in 12, 24, and 48 volts, and can be utilized in off-grid as well as hybrid PV systems.

Off-Grid Solar Power Systems (Standalone Solar Power Systems)

Off-grid solar systems allow you to be energy-independent by generating your own power from the sun. You will not rely on the electrical grid to supplement your energy needs. An efficient solar power system coupled with an excellent battery storage system is required to be installed in your home. An off-grid PV system allows you to have complete control of your energy production at the site of power consumption. Installing a standalone system requires more installation space and higher upfront costs. Over recent years, PV equipment has become more efficient and cost-effective. So, not only do you power your home, but you can also install it in your RVs and cabins. Unlike the other types of solar power systems, the off-grid system doesn't connect to other external sources of power, so utilizing a power generator might be extremely beneficial for cold, cloudy winter days. An off-grid solar system has components that allow you to generate power, store it, and supply it onsite.

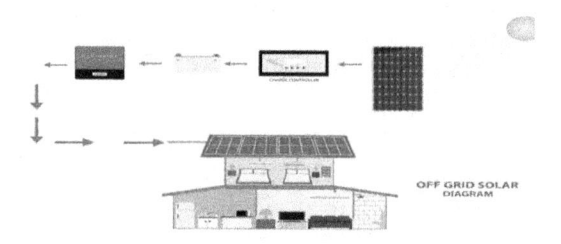

OFF GRID SOLAR DIAGRAM

Uses of Off-Grid Solar Power Systems

The scalability of solar energy is one of the biggest reasons why it is the most preferred source of independent energy. A combination of various system components increases the flexibility and uses of solar power.

It ranges from a simple PV system to an extensive power system that will power a factory. Some of the common uses of off-grid solar power include:

• Providing energy needs for portable devices like cell phones or tablets.

• Providing energy-efficient homes.

• Generating electricity to power RV appliances and small cabins.

Advantages of Off-Grid Solar Power Systems

• **Available in remote areas:** it is a suitable option for those who live in remote areas where there is no access to grid power.

• **Self-sufficient:** designing an efficient solar system ensures you meet all your energy demands. Additionally, if there is a grid failure, it will not affect your power supply.

Disadvantages of Off-Grid Power Systems

• **Expensive:** an additional cost of installing the batteries and other components increases the installation cost when compared to grid-tied systems.

Sometimes, you need to install generators as an alternative source to add to the initial expenses.

- **High-maintenance:** additional components used in the installation require frequent maintenance.

For example, some lead acid batteries require regular refilling.

- **Battery replacement:** generally, battery banks last for around seven to ten years, and you have to replace them. This increases the cost of maintaining your solar system.

Components of an Off-Grid Solar Power System

For your system to function efficiently, you must select components that meet your energy demands. The main equipment required includes:

Solar Panels

Solar panels are the key element in your installation process. When buying solar panels, you should evaluate them in terms of cost, efficiency, technology type, and the warranty. The most commonly used solar panels for residential uses include monocrystalline panels and polycrystalline panels. Monocrystalline panels are the most expensive, yet efficient panels on the market.

The number and size of solar panels you buy depends on:

- Your household appliances and energy requirements.
- Efficiency ratings.
- Size of your roof area.

- Peak sunlight exposure in your specific location.

Solar Charge Controller

The charge controller allows you to use DC power directly, as well as regulate the DC power generated by solar panels to charge the batteries. It is connected between the panels and the battery to maintain proper charging of the battery. It regulates the voltage and current flowing to the battery to prevent overcharging of the battery.

Solar Inverter

An off-grid inverter converts the DC power generated by panels to the AC power that's required to run appliances in your home.

Battery Bank

A backup battery stores the excess energy for later use. The battery bank enables your PV system to continue operating with minimum sun exposure.

Mounting/Racking System

A racking system's function is to secure the panel(s) on the roof since you can't directly mount the panels on your house's roof.

Hybrid System versus Off-Grid System

If you go off-grid, you will only have the backup battery to supply your energy needs. A hybrid system will allow you to draw power from the battery or the grid.

An off-grid PV system allows you to have complete control of your power generation; however, you have to properly manage your power consumption so that your batteries don't run out.

If you have a hybrid system, however, you don't have to worry about your energy consumption. Even if you overuse your battery power, you can still get more power from the grid. This system is suitable for you if you have enough funds and don't want to take the chance of your power running out. So, if your energy demand is high and you need a constant power supply, you can go for this option.

Chapter Summary

Setting up a grid-tied system without batteries is less complicated because it is less expensive, and there is no need to incur the maintenance cost of the batteries. Although it is more efficient, a grid-tied system is not an independent power supply source. If there is an interruption of the main power supply from the grid, you will not have lights even if there is enough sunlight. Alternatively, you can set up a grid-tied system with a battery backup to provide you with extra energy storage for use during off-peak periods. This system allows you to draw power from the battery bank or from the power grid.

An off-grid PV system is a standalone setup that uses its own photovoltaic panels and battery to store solar energy. The system is not connected to any local utility company or a power grid, as is the case with grid-tied systems. Off-grid systems can work well in remote or rural areas where there are no power lines and the cost of extending a power line from the grid is considerably high.

In the next chapter, we will discuss in detail the most well-known solar component: solar panels. In fact, you will learn the practical steps for setting up a PV system.

DIY SOLAR POWER FOR BEGINNERS

CHAPTER FIVE: Solar Panels (PV Modules)

The photovoltaic (PV) system utilizes a sustainable source of energy and it's more affordable than ever before. PV effects cause solar panels to convert the light energy from the sun into electrical energy. When sun rays shine on the surface of the panels, they are absorbed, reflected, or even pass through the cells to generate electricity.

PV Module and PV Cell

Since solar panels utilize photovoltaic (PV) technology, they're referred to as PV modules. PV modules are composed of a certain number of PV cells depending on their power-generating capacity.

PV modules are made of PV cell circuits enclosed in an environmentally protective laminate; they act as the building blocks for the PV solar system. PV cells act as a semiconductor material (silicon) that converts light energy into DC electricity.

A group of PV modules is wired together to form a PV array. PV modules may be wired in parallel, series or both to deliver enough current and voltage required to run appliances.

The front surface of the PV modules is covered with a transparent material, such as tempered glass, and a waterproof material at the back.

Also, a weatherproof material covers the edges of the module, while the aluminum frame holds all the components together to form a mountable unit. Each PV module has a junction box or a wire lead at the back to allow you to connect the modules to other solar system components.

How PV Modules Generate Electricity

A single PV cell can generate one to two watts of power. The PV module boosts the power output by connecting a number of PV cells together. When PV cells absorb light energy from the sun, the photons produced are absorbed into the semiconductor material. This forces electrons to flow through silicon in the form of an electrical current. The cells only act as an electron pump since they don't have storage capacity. The number of electrons flowing depends on the photons produced. Exposing bigger and more efficient PV cells to intense sunlight will increase the flow of electrons. The more electrons, the more electricity is generated.

Types of PV Modules

There are three common manufacturing techniques for PV modules. These technologies result in the production of solar panels that vary in terms of appearance, cost, performance, and method of installation.

Single Crystalline/Monocrystalline Silicon

This is a form of crystalline silicon solar panel with a high energy efficiency of 17% to 20%, providing an efficient solar energy production technology.

These PV modules are space-efficient, expensive, and have a high energy-bearing capacity. Due to the high lifetime value of these panels, they're optimized for commercial and residential purposes. The panels have rounded edges, a uniformly darker blue color when compared to multi-crystalline modules, and are less affected by high-temperature changes.

Multi-Crystalline/Polycrystalline Silicon

This works similarly to the single crystalline, but has a lower conversion efficiency of between 13% and 16%. The cells of these modules are made of several silicon crystals, making them cost-effective. They are easy to identify due to the square shape of the panels, and they have a blue speckled look. The panels have a slightly shorter lifespan and are more affected by high temperatures.

Amorphous Silicon (A-Si)/Thin-Film PV Module

This thin-film PV module has a lower light absorption rate than the crystalline silicon modules. These modules have the lowest efficiency of 10%, but are more installation-friendly. These modules have a thickness of a few nanometers to micrometers, hence the name "thin-film". They are the most portable, flexible, easy-to-install, and lightweight panels on the market.

If you're on a low budget, you may want to select this option. Many people prefer these panels because of their flexibility and the fact that they're less affected by high temperatures. The lightweight nature of thin-film panels makes them easy to install on any surface, such as glass, metal, and plastic.

They occupy the least possible space, making them suitable for installation on an RV or van.

Solar Panel Specification Sheet

Solar panel specification sheets provide detailed information about how to operate the panels and how to configure the PV system. You will also know the power production capacity and the efficiency, how temperature changes affect the operation of the panels, and information on the dimensions of the panels. This information allows you to analyze the performance of the panels accurately.

You can review this information at the back of the panels or review what your installer recommends so that you know what you're buying. You can find the datasheet information in PDF format under the product, from the support, or download it from the manufacturer's website.

There are a number of terms and ratings you need to understand to avoid confusion when reading your solar panel's datasheet. The main ratings are reflected in the following datasheet: You need to learn theses terms to understand and utilize them for your system applications.

Standard Test Conditions (STC)

STC measures the performance of solar panels based on a set of criteria. Typically, the voltage and current generated by the panels vary based on the intensity of light and temperature changes, among other criteria.

Specifications

Electrical performance at STC

Maximum power	125 W
Maximum power current (Impp)	7.18 A
Maximum power voltage (Vmpp)	17.4 V
Short circuit current (Isc)	7.14 A
Open circuit voltage (Voc)	20.92 V
Maximum system voltage	600 V
Temperature coefficient of Isc	0.0045 A/^{0}C
Temperature coefficient of Voc	-0.085 V/^{0}C
Efficiency	19.8%

Electrical performance at NOCT

Maximum power	90 W
Maximum power current (Impp)	5.96 A
Maximum power voltage (Vmpp)	15.1 V
Short circuit current (Isc)	6.124 A
Open circuit voltage (Voc)	19.5 V
Efficiency	18.4%

The standard test conditions for testing all solar panels include an ambient temperature of 0-2^{0}C, cell temperature of 25^{0}C (77^{0}F), an atmospheric density of 1.5, and a light intensity of 1000 watts per square meter. For instance, for a 370-watt Sunpower solar panel, the module's maximum generated power would be 370 watts only under standard test conditions.

Normal Operating Cell Temperature (NOCT)

There are periods when the normal temperature is not 77 ^{0}F, especially during the summer months. Extreme temperature ranges can affect the operational parameters of your solar panels. NOCT considers real-world conditions and provides actual power ratings of your solar system.

Instead of using 1000 watts, NOCT uses 800 watts/square meter, which represents a sunny day's irradiance, cell temperature of 45^{0}C, and air temperatures of 20^{0}C (68^{0}F).

As explained, panel specifications are measured and mentioned in both STC and NOCT to help you figure out which and how many panels are required to meet your specific needs.

Current-Voltage (I-V) Curve

I-V curve is an important feature that provides performance information on an open circuit voltage, short circuit current, maximum rated power, maximum current, maximum voltage, and module's efficiency.

These functions are important in designing, utilizing, testing, maintaining, and controlling PV systems.

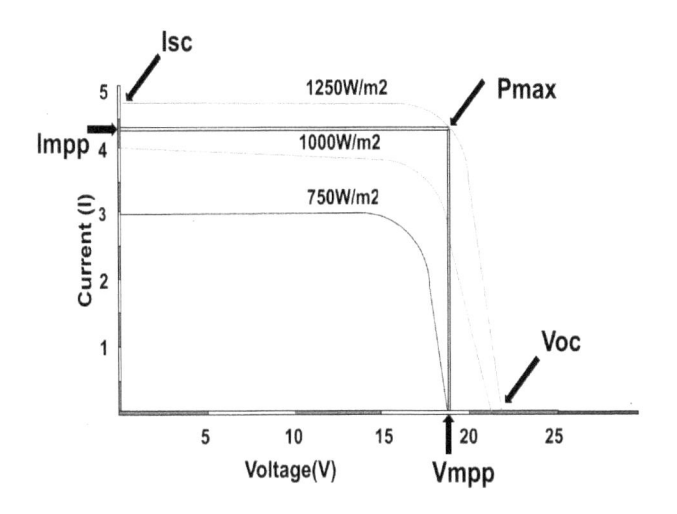

I-V Curve

Using the I-V curve is the simplest way to illustrate the relationship between the current flowing through PV cells and the voltage applied across them. Understanding this curve helps you determine the efficiency and output performance of the panels.

The intensity of sun rays that hit the PV cells controls the generated current, while the increase of the PV cell's temperature reduces the voltage.

When the PV cells are not connected to any load (open circuited), the current will be zero and the voltage will be at its maximum (open circuit voltage).

On the other hand, if the cells are short-circuited (both positive and negative leads are connected to each other), the voltage across the cell is at zero while the current is at its maximum (short circuit current).

Typically, in the I-V curve, the current and voltage figures range from the maximum current (Isc or short circuit current) with an output of zero volts to the current of zero and maximum voltage at Voc (open circuit voltage).

Currently, many PV modules are equipped with a maximum power point tracker (MPPT) that utilizes the I-V characteristic to assess the performance of the PV module.

Short Circuit Current (Isc)

The short circuit current (Isc) indicates the amount of current (how many amps) produced by the solar panels when not connected to a load (top left of the curve). In this condition, the solar panel's positive and negative terminals are directly connected to each other. You can use an ammeter to read Isc across the positive and the negative leads. This indicates the highest amount of current generated by your panels under STC or NOCT. Generally, you need the Isc to determine the maximum amps the conductors, charge controller, and inverter should be able to handle within your setup.

Open Circuit Voltage (Voc)

Voc describes the voltage output of solar panels when there is no load connected. You can use a voltmeter to measure the positive and negative leads to read the open circuit voltage.

Since solar panels are not yet connected to any other device, there is no load on it and no current flowing through them. Voc is considered as the highest amount of voltage generated.

It is very important to note the Voc figure because it represents the maximum voltage produced by solar panels. As observed in I-V curve, Voc is located at the bottom right, where the current shows zero. It helps you determine how many solar panels you need to connect in series to the charge controller or the inverter. Early morning, when the sun first shows up and the panels are at their coolest condition, is the right time to measure the Voc number since most household's heavy loads are switched off. Because your system fuses and circuit breakers will only protect against overcurrent and not overvoltage, exposing your electronic devices to higher voltage levels will damage them.

Maximum Power Point (Pmax)

Pmax represents the maximum power output of the solar panels. As you can observe in the previous I-V curve, the knee of the curve, where both current and voltage are high enough to generate the highest power (wattage) possible, corresponds to the maximum power point or, MPP: Watts = Volts * Amps

If you're using a Maximum Power Point Tracking (MPPT) charge controller, the knee of the curve (Pmax) is where MPPT devices try to maintain the volts and the amps to maximize power output. If voltage is increased, for example, because of lower temperatures in the morning, the current is decreased accordingly, and, as a result, the generated power would be less than the maximum power point. Therefore, the wattage listed for the solar panels refers to the maximum power point: Pmax = Vmpp * Impp.

Maximum Power Point Voltage (Vmpp)

This is the voltage of your solar panel when the power output is at its highest. It is the voltage your MPPT charge controller or grid-tied inverter displays at noon when your panels are converting the intense sunlight to maximum power (Pmax) at STC or NOCT.

Maximum Power Point Current (Impp)

Impp is the current generated when the power output of the panels is at its highest. It is the amperage displayed by the MPPT charge controller while the panels are generating the maximum power (Pmax) at STC or NOCT.

Nominal Voltage

Nominal voltage is not the actual voltage measured from solar panels. It is just a category and shouldn't be confused with the actual voltage output of the panels. Nominal voltage helps you match your solar equipment; that is, know what equipment goes together. For example, a nominal solar panel of 12 V with a Voc of 22 V and Vmp of 17 V should only be connected to a 12-volt charge controller and inverter to charge a 12-volt battery backup. If you decide to utilize a 24-V battery and charge controller, you have to connect two (strings of) 12-V solar panels in series to raise your system's nominal voltage to 24 volts.

Solar Panel Efficiency

The efficiency of solar panels depends on how much incoming sunlight generates electricity.

Higher efficiency means having fewer solar panels to produce the same power output and consequently, less needed roof space and racking equipment along with reduced labor costs.

The only tradeoff of having high-efficiency panels is the higher cost of each module, which will be compensated by the fewer number of panels needed.

In case you decide to utilize one micro-inverter or optimizer per panel, the high-efficiency panels are an advantage. Since you need fewer panels, fewer micro-inverters are required, hence the lower upfront cost.

You will find the efficiency of panels listed on the datasheet as a percentage. As solar technology keeps changing, there are several PV modules with higher efficiency percentage than 20% on the market, like SunPower SPR-X22-360 with 22.2% efficiency, and LG 375W with 21.7% efficiency.

Monocrystalline panels have the highest efficiency and can absorb light energy both at the front and from the rear.

If it's your first time buying the panels, you can lay out your system with both high-efficiency panels and low-efficiency ones to analyze the result and know which meets your energy demands as per your budget.

You can also ask your solar installer to do so. By performing some simple calculations that will be explained in this chapter, you can compare as many different panels as you want.

Temperature Coefficient of Power

Just like any other semiconductor device, PV cells are sensitive to extreme temperature changes. An increase in temperature affects the open circuit voltage of your panels. Temperature coefficient is an indicator of how temperature affects PV array's power output. If the temperature increases, the power output decreases.

Most manufacturers provide coefficient information on their brochures. If the value of the temperature coefficient of power is not indicated on the brochure, you might see a graph that shows the normalized performance plotted against the PV cell temperature or a table showing Isc and Voc for different temperatures. Below you can see a temperature-related graph:

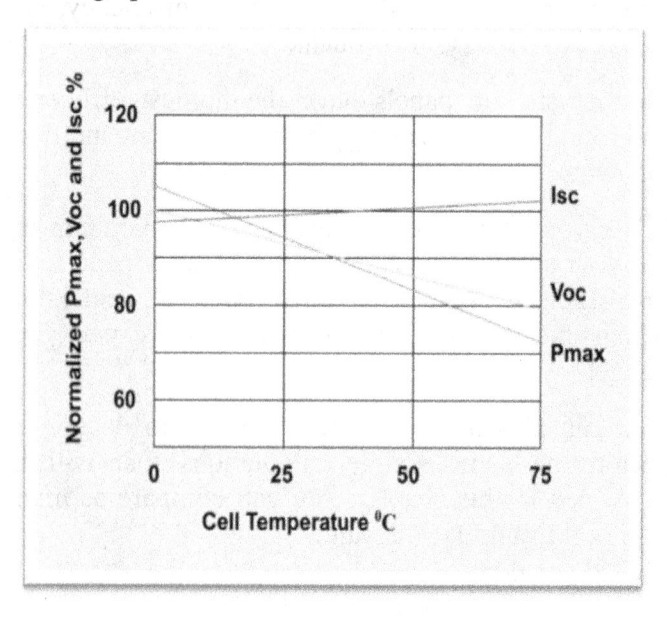

In this example, the sloping power line (Pmax line) illustrates the negative effect of higher temperatures on the generated power. You can also see the short circuit current and a normalized open circuit voltage.

Generally, the PV module's power is rated at 25^0C. Any increase in temperature to above 25^0C will result in a power loss of about 1% for every 2^0C increase. In many sunny countries, temperature can rise up to 50^0C.

Voc and Isc Temperature Coefficients

Temperatures that are lower or higher than 25^0C have a significant impact on the open circuit voltage (Voc) and short circuit current (Isc) of the solar panels.

Low temperatures will result in higher Voc than test conditions. Unlike voltage, the short circuit current goes significantly high in higher temperatures.

Each PV cell has its own temperature coefficient for Voc and Isc, and you can easily determine the Voc and Isc of your system according to your local temperature.

As the previous example specification sheet implies, the temperature coefficient for Voc is $-0.085/^0C$, and the Voc under NOCT is 19.5 volts. It means that when the ambient temperature is 5^0C, then:

$5^0C - 25^0C = -20^0C$ difference

$-20^0C * -0.085 = 1.7V$ increase in Voc

19.5 volts + 1.7 = 21.2 volts, Voc at 5^0C

Assuming that the highest local temperature reaches 45^0C, let's calculate the Isc at 45^0C:

45^0C - 25^0C = 20^0C temperature difference

20^0C * 0.0045 (temperature coefficient for Isc) = 0.09 amps increase in Isc

6.12 amps + 0.09 amps = 6.21 amps Isc at 45^0C

Depending on your weather conditions, you need to find the adjusted figures to design the right size of other components, such as batteries, charge controllers and conductors.

PV Module and Shading Effects

PV cells in the solar modules are sensitive to radiant light energy emitted by the sun. In order to maximize the energy generated by the modules, you have to first understand the PV cell's wiring inside the modules.

Additionally, it's important to identify factors that influence the performance of the cells.

To obtain the desired voltage output from the PV module, the 0.5-volt cells are connected in a series; however, cells wired in series may experience some issues when one of the cells cannot generate power due to shading.

The amount of electricity generated will be lower than that of unshielded cells. As you may recall the Christmas light string example, if one light does not work, the whole string will shut off.

The shaded cell dictates the total electricity of the module. As a result, it will lead to power loss. When some cells are shaded, a higher amount of electrical current generated by the unshaded cells passes through the shaded ones; therefore, the affected cells might act as a load and increase the temperature. This shading effect will lead to a "hot spot" problem. The hot spot is where the increased temperature will result in a change of cell characteristic and will lead to irreversible damage of the PV cells within the module, such as having a cracked glass or melted cells.

How to Overcome Shading

To prevent hot spot problems, as well as improve power output from shaded cells, PV modules are equipped with bypass diodes. A bypass (internal) diode will block the shaded cells so that no current will pass through them. The diode is either integrated into the module itself or installed in the module junction box. Each module has three strings of series cells: left, middle, and right strings, each of which has one diode. For example, if one or two rows of cells are shaded, their diode blocks them and bypasses the current so the unshaded cells are not affected.

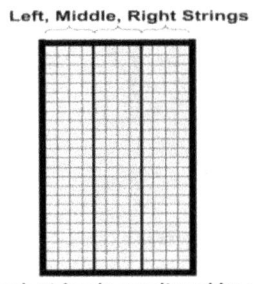

Assuming you have a 60-cell module, you will have three strings of twenty cells wired in series, each protected with a single diode. The three strings are connected in parallel with each other.

Shading of the PV Array/String

A PV string consists of one or several modules connected in series. Do not confuse PV string with the strings inside each solar module (panel). Wiring the modules in series increases the voltage output while the current will remain the same.

A solar array, however, is usually composed of one or several strings containing the same number of panels. These PV strings are connected in parallel, so the current will be scaled based on the number of strings while the voltage will be the same as that of each string.

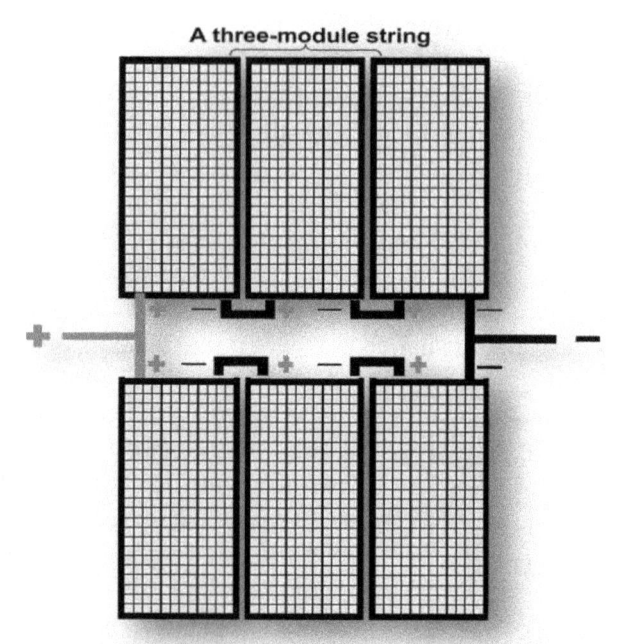

A three-module string

An array composed of two identical strings

The intensity of sunlight in a particular area rarely affects the voltage output of your system; however, the current generated by the PV module is dependent on the amount of sunlight the modules are receiving.

Shading of individual modules affects the performance of your PV array. A reduced generated power by one shaded module will result in a reduction in the rest of the modules connected in series in the same string.

The most efficient method to avoid the shading effect in PV string is the use of Module Level Power Electronics (MLPEs).

These include power (DC) optimizers and micro-inverters, which can be connected to an individual PV module to increase its performance under shading conditions.

By using MLPEs, maximum power point tracking is done at the module level.

DC Optimizers

The DC optimizer adjusts the voltage and current flowing through each PV module so as to maintain a maximum power output without affecting the performance of other modules.

For instance, if a shaded module generates lower current output, the DC optimizer boosts its current output and reduces the voltage by the same amount.

This ensures the flowing current matches the current flowing through the unshaded modules connected in series in that string.

Micro-Inverters

Instead of using a single central inverter, you can have each panel connected to a small inverter. That's because each micro-inverter has MPPT and each of the panels will operate at its maximum power point without affecting the output of other panels. Unlike power optimizers, micro-inverters convert the generated DC power to usable AC electricity on the roof.

Irradiance and PV Performance

Just as an increase in temperature reduces power output, irradiance also influences PV performance. Lower exposure to sunlight will result in reduced production of current, and consequently, it can affect the power output.

Irradiance is defined as the instantaneous amount of solar power per unit area that hits the surface at a particular angle. The irradiance unit of measurement is watts per square meter (W/m^2) or kilowatt per square meter (kW/m^2). As our planet rotates around the sun, the distance between them always changes. As a result, the amount of solar irradiance varies from one period to the next.

According to NASA, the solar constant or the average irradiance value on a flat surface perpendicular to the sun is 1370 W/m^2; however, in reality, irradiance measured on the Earth's surface is lower than this. This is because the climatic conditions affect the scattering and reflection of sunlight as it penetrates through the atmosphere and reaches the Earth's surface.

Annual climatic conditions, such as time of the year, temperature variation, cloudiness, and the angle at which sun rays strike the Earth's surface, have an impact on the amount of solar irradiance in a particular area. During sunny days, especially in summer months, solar irradiance is higher than that of winter months.

Measurement of Solar Irradiance

When designing your PV system, you have to estimate the amount of sunlight available in your location at any given time. Solar radiation values are mostly dependent on your location and the local weather. Direct radiation measurements are released periodically as the Global Horizontal Irradiance (GHI). You can download the global horizontal irradiance maps here, to identify the amount of sun exposure in your own location.

These maps indicate annual and monthly average geospatial data stored in the National Solar Radiation Database (NSRDB) on a Physical Solar Model (PSM). The PSM mostly covers the United States.

Countries, like the United States, receive more solar energy because they lie in the middle of latitude where the sun is nearly overhead. They receive more sunlight in the summer months because days are longer. During the shorter days of winter months, the sun's rays are slanted. Further, the Earth's rotation results in hourly variation of sunlight. At noon, the sun is at its highest point. In the early mornings and afternoons, it is low in the sky.

Peak Sun Hours

Peak sun hour is defined as an hour of sunlight with an average intensity of 1000 watt/m^2 (around 10.5 square feet). Before switching to solar power, you have to determine how many peak hours you receive in your location. This will give you an accurate estimate of how much power you will generate with your system.

The peak sun hours per day influence the number of panels needed to cover your power consumption. These peak hours vary from daylight hours. On average, the panels get exposed to sunlight for seven hours a day; however, the average peak sun hours in a day are generally four to five in the United States, and the sun reaches its peak point at noon.

The number of peak hours per day increases as you get near the equator and during the summer months.

If you live in Phoenix, for example, you can experience higher peak hours than someone living in Seattle.

You can use the Renewable Resource Data Center that provides you with peak sun hours from state to state.

Alternatively, you can use the peak sun hours map by scanning the code below to estimate the high, low, and average peak sun hours of your specific city.

Keep this number to use for calculating your system power capacity.

To maximize your solar output, you have to install your panels in such a way that they receive direct sunlight.

The panels should face south and away from any shading from trees. Remember, even shading in a small area of the panel can affect the power output since the solar cells are connected in series.

How Does Irradiance Affects PV Output?

If all of the parameters remain constant, a higher irradiance will result in higher current output and generate more power.

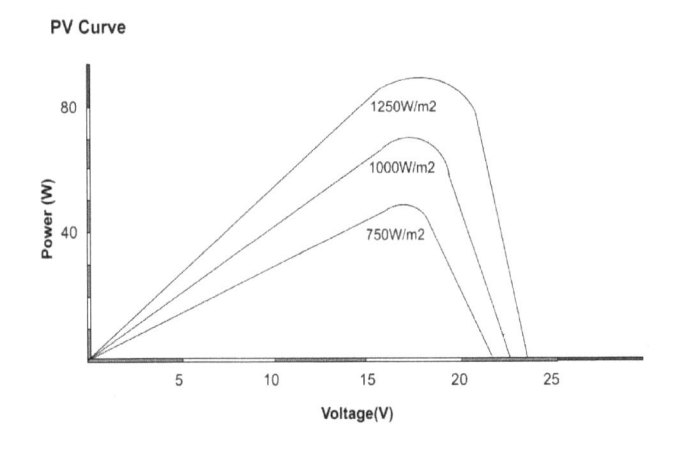

PV Curve

The above curve shows the relationship between voltage and power of PV modules at different amounts of irradiance. When irradiance increases, the module produces a higher electrical current, and consequently, more power, as indicated in the vertical axis.

You can also observe the relationship between voltage and power at different irradiance levels. An increase in irradiance will result in a high generation of power, and the high peaks on the curve represent output power. This relationship between current and power in the PV module is expressed as:

$$\frac{G1}{G2} = \frac{I2}{I1} = \frac{P2}{P1}$$

Where, G_1 and G_2 are irradiance levels measured in W/m^2, I_1 and I_2 represent current (measured in amps), and P_1 and P_2 represent the power at different irradiance levels (measured in watts).

Tilt and Orientation of the PV Array

When mounting solar panels, you should ask yourself the following questions:

1. At what tilt you should mount the modules.

2. Whether to use portrait or landscape orientation.

Tilt Angle

Tilt angle is defined as the vertical elevation angle at which modules are mounted on the roof. When installing the modules; you can set the orientation, or direction of the tilt to optimize the panel's performance. If you have a pitched roof, however, the preferred module's tilt angle will be the same as the tilt of the roof.

In a flat roof, the tilt is 0 degrees, while for a vertical wall-mount module; the tilt angle is 90 degrees. The amount of energy generated depends on the tilt angle.

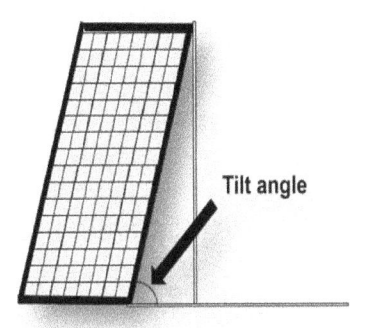

Tilt angle

Tilt Angle and Latitude

At noon, the sun is not always above us. By tilting the panels slightly on the south if you live within the Northern Hemisphere, you can maximize the annual output. The tilt angle depends on the latitude; the further you move from the equator, the higher the tilt angle.

Electricity generation will be at its peak in the afternoon and in the early evening if the panels are facing toward the west. In this case, panels can give the maximum yields over panels facing toward the south because they tilt toward the setting sun.

If the PV array has exactly vertical exposure to sunlight throughout the day, the output from the array will be at its highest; however, the sun moves throughout the day, making it impossible to face direct sunlight unless you use a dual tracking system, such as (2-axis solar tracker).

This tracker ensures the array tilts at an optimal angle throughout the day and in different seasons to maximize the output.

Alternatively, you can install a single-axis tracking system on the tilted arrays to enable the panels to rotate to face the sun as it moves from east to west. This increases the output during the early and late daylight hours. Installing this system will increase your initial cost by around 40% to 50% while raising your power output by 30% to 35%. Since adding solar trackers will significantly increase your upfront costs, most homeowners ignore this option and prefer a fixed mounting.

Tilt Angle for Flat Roofs and Ground-Mounted Systems

As a rule of thumb, your mounted solar modules' tilt angle should match the latitude of the selected location.

Panels mounted at angles between 30^0 and 45^0 work well in most locations.

Additionally, lower tilt angles result in higher energy production during the summer months, while a high tilt angle can scale up power generation during the winter months. This is apparently due to sun's changing position in the sky throughout the year.

By using the System Advisor Model (SAM), a free software model provided by National Renewable Energy Laboratory (NREL), you can calculate the output from different tilt angles.

By scanning the code below, you can check their website:

The following example explains how tilt angle affects the amount of power generation. Assuming you have a 3.4 kW PV system with sixteen, HIT-215A panels located in Portland, OR (45^0 north), the annual production output is calculated for different tilt angles:

Array Tilt Angle (Degrees)	**Portland, OR (45^0 North)**	
	Annual Production (kWh)	Delta (%)
0 (flat)	3624	0%
10	4019	9%
20	4239	15%
30	4355	18%
40	4368	18%
50	4279	16%
90 (vertical)	2967	-20%

As you can observe, the tilt angle of 40 degrees indicates the greatest amount of energy output. Cities in the northern parts have a higher positive effect (delta) of tilt angle than those near the equator, meaning that the modules are highly required to be angled equal to the location's latitude in order to generate the largest possible amount of power.

Generally, if the panels are installed facing directly east or directly west, they will produce less electricity than if they're facing south.

Tilt angle and Wind

Wind direction, wind speed, and the inclination of PV arrays influence the efficiency of solar energy production, especially in hot seasons. Since high temperatures affect the performance of solar panels, the wind cools down the panels. As a result, the panel's energy output will increase. Furthermore the inclining angle of the panels determines how incoming airflow is distributed under the panels. The speed with which air flows between or under the panels determines the panels' cooling effect.

Tilt Angle for Small Cabins and Portable Systems

There is no fixed tilt angle that provides you with maximum power because the sun moves across the sky during the day and alters throughout the year. The amount of energy generated depends on the season and time of day.

Therefore, to calculate an optimal tilt angle, you have to determine the direction where your panels should face. This depends on:

- Where you live.
- Time of the year in which you need the most energy.

As a rule of thumb, you should mount the panels more vertically during the winter in order to get the most of the winter sun. And the lower you tilt the panels during summer, the higher the energy output will be. You can calculate the optimal tilt angle for small cabins based on your latitude.

To get an optimum angle, you should add 15^0 to your latitude during winter and subtract 15^0 during summer. If you live within latitude of 42^0, your tilt angle will be $42 + 15 = 57^0$ during winter and 27^0 during summer.

Alternatively, you can calculate the optimum tilt angle during winter months by taking latitude multiplied by 0.9 and then add 29^0. In the above location with a latitude of 42 degrees, the tilt angle will be $(42 * 0.9) + 29 = 66.8^0$.

This method is more effective in tapping the midday sun than the previous method and gives you better results.

During the summer months, the optimal tilt angle can be calculated as follows:

Latitude multiplied by 0.9 and subtracted by 23.5^0:

$(42 * 0.9) - 23.5 = 14.3^0$

During spring and fall, just subtract 2.5^0 from the local latitude.

Azimuth Angle

The azimuth angle is the horizontal orientation of PV panels in relation to the equator. Finding the right tilt and azimuth angle helps you get maximum power output from your solar array.

To get optimal output from your array, you should mount them facing south if you live in the Northern Hemisphere, but if you live in the Southern Hemisphere, the panels should face north.

If you're relying on your compass to mount the panels on the ground, you may find that it's not giving you an accurate direction. And, depending on your location, the compass can give an inaccurate reading of more than 25^0. If you live in the US, for instance, you should mount the panels facing the true south instead of relying on a compass that points to the magnetic south.

This is because the magnetic forces in the Earth's core tend to pull the compass needle away from true south or true north.

The difference between true north and magnetic north, as shown in your compass reading, is known as *magnetic declination*. This means extra compensation degrees from the compass are required to obtain true north.

A positive figure indicates the declination of panels to the east of the compass reading to form a true north. A negative figure indicates a western declination, and this represents true north to the west of the compass reading.

Therefore, to find the azimuth angle for your panels, you first have to find the magnetic declination using either NOAA.gov's calculator or the available online charts. Based on your location, you can adjust the magnetic declination value to position the panels at an angle that gives you the maximum yield.

If you live in the Northern Hemisphere:

- Rotate the panels to face east if the magnetic declination (positive) points to the east.

- Rotate the panels to face west if the magnetic declination (negative) points to the west.

If you live in the Southern Hemisphere:

- Rotate panels to face west if the magnetic declination value (positive) points to the east.

- Rotate panels to east if the magnetic declination (negative) points to the west.

For example, if you live in San Diego, which is in the Northern Hemisphere, your magnetic declination is 11 degrees east; therefore, you should find the magnetic south and adjust the panels 11 degrees to the east.

But if you live in Chile, which has a magnetic declination of 11 degrees east in the Southern Hemisphere, you have to adjust the panels to face north. So, you should adjust to 11 degrees west to find the ideal azimuth angle.

Making these adjustments ensures your panels face directly to the equator and maximizes exposure of panels to the sunlight, giving you optimal power output.

For home roof installations, however, the best option is a south facing roof in the Northern Hemisphere because the panels' horizontal angle follows the roof's orientation.

PV Module Orientation

The array orientation can either be portrait or landscape.

Portrait orientation is the vertical layout of modules in which the short side of the module is mounted parallel to the ground.

The landscape is the horizontal layout in which the long side of the module is parallel to the ground.

Mounting panels in landscape orientation requires more railing materials when compared to having vertical rows of panels.

In most cases, the size of solar panels makes them suitable for vertical (portrait) installation.

Whether to choose portrait or landscape depends on the number of panels you can install on your roof, the tilt angle, and the shape of the mounting area.

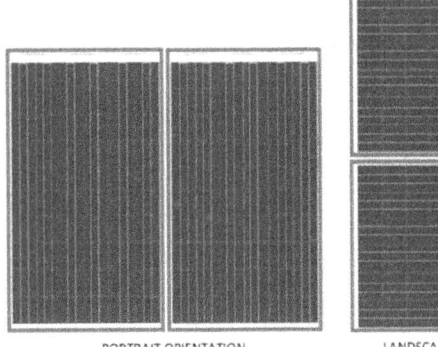

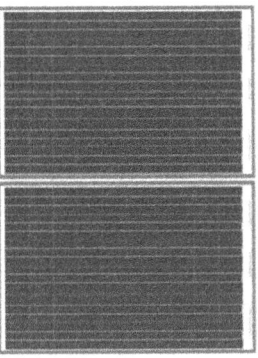

PORTRAIT ORIENTATION   LANDSCAPE ORIENTATION

Therefore, your decision on how to mount the panels is influenced by your location, shading as well as snow. If you live in an area that experiences snow, it is best to mount the panels in landscape (horizontally) because the string of cells in the module will run lengthwise, leading to more energy production.

By installing them horizontally, snow will settle at the bottom of the panels, thus only blocking one string of cells in one panel while leaving the rest of the cells clear from snow and exposing them to sufficient sunlight.

But if you mount panels vertically, snow will slide downward to the bottom of each panel. This will partially block all strings of cells on solar modules, reducing the power output due to the partial shading effect.

In addition, the amount of sunlight exposure on the roof can also impact the direction you mount the panels. If one or more cells are shaded, current will not flow through that cell. This shading will affect the total energy output generated by the panels.

Solar Panel Sizing

Investing in solar systems is a smart move for every homeowner; however, to make the most of your system, you have to know how to choose the correct size that covers your energy needs.

Before you begin to size your PV system, ask yourself the following questions:

- Budget constraints: do you intend to build a PV system that fits your target budget?

- Energy production: do you intend to build a system that offsets a certain percentage of your monthly energy consumption?

- Special constraints: do you want to design a system that is space-efficient?

Once you decide on the type of PV system to design, you should consider the following factors as well:

- Level of sun exposure in your location.
- PV array orientation and tilt angle.
- Future expansion plans.
- Ratings on PV efficiency.
- Degradation of the system over time.

After establishing your energy needs and design approach, you should follow these steps to estimate your load and, consequently, the size of your system including the number and size of PV modules as well as the number and capacity of your system's battery bank if indicated.

PV Sizing for Grid-Tied Systems

Method 1:

Step1. Estimate your Energy (Calculate Your kWh Usage)

The first step to sizing your solar system is to determine your average daily power consumption (kWh). This will help you in knowing how many panels you need to install in your home.

Don't forget that following the exact steps mentioned here will allow you to have a more accurate estimate of what your PV system will look like once everything is installed.

Start by gathering your electricity usage (kilowatt-hours) based on your electricity bills from the utility company to determine your consumption for the last twelve months. From this, you can identify your peaks in electricity usage throughout the year.

Mostly, your energy usage spikes during the winter and summer months because of the heavy use of the heating and cooling systems.

In the summer months, your grid-tied system tends to produce more electricity due to peak sun exposure.

In the winter months, however, the system will generate less power due to lower sun exposure.

The following table represents an electricity usage list. By preparing a list like this, you will have a precise estimate of your monthly electricity consumption:

Electricity Consumption History	KWH
June 14	850
July 13	1123
Aug 13	1148
Sept 13	1058
Oct 13	1127
Nov 13	834
Dec 13	945
Jan 14	869
Feb 15	705
Mar 14	682
Apr 14	679
May 14	680

From your annual power consumption bill, you can get your average daily electricity usage. Add up your power consumption for twelve months. In the example above, the annual consumption is 10700 kWh. Then divide the total number by 365 days to get the average daily power consumption:

(10700 / 365) = 29.31 kWh, average daily consumption

Step2. Determine Peak Sun Hours

Your local peak sun hours depend on your location and climate in the area.

You have to determine the peak hours of sunlight per day in order to get the most of your solar power.

Identify peak sun hours in your geographical area to estimate how much energy your panels can produce during the peak hours. You can use the previously-mentioned sun hours map chart to get the average peak hours for your city. Assuming you live in Arizona, your array will experiences 5.5 sun hours per day.

Using annual average daily sun hours will help you roughly estimate your average daily power generation; however, it does not reflect the actual power generation potential of your system on sunny, summer days and cloudy, winter days. Instead, you can consider the winter daily average sun hour to be more conservative.

Step3. Calculate the Panels Output

To obtain the power output of your PV system, take your daily power consumption and divide it by peak sun hours. In this case, let's calculate how much energy your panels generate each hour.

Start by multiplying your hourly power usage by 1000 to convert your power consumption into watts. Then divide it by the number of daily peak hours.

29.72 kWh * 1000 = 29720 watts

Solar panels output = daily power consumption (kWh) / average peak sun hours

Assuming you live in Arizona, which experiences 5.5 peak sun hours per day, your array's output per hour will be:

29720 watts / 5.5 sun hours = 5403.6 (rounded to 5404 watts).

PV setups do experience system losses from the solar inverter, connected cables, and others that amount to 25% of the system's total power; therefore the actual size of the system is derived by adding 25% to the solar array's output:

5404 watts * 1.25 = 6755 watts

Based on your roof size, location, peak sun hours, and grid reliability, you can decide what percent of power consumption to cover by your panels. For the example above, we decided to cover 75% of the daily consumption; however, most homeowners consider 50% to 60% when they first install a PV system:

5404 watts * 75% = 4053 watts desired output

Step4. Calculate the Size of Your PV Pystem

Lastly, divide the solar array's output by energy rating for each individual panel. Since panels are rated based on individual consumption, most of the panels are in the range of between 275 and 380 watts; if you choose a 360W High-Efficiency LG Solar Panel, we can refer to the datasheet and figure out the Pmax under NOCT is 325W; therefore, the number of panels needed for your system will be:

4053 watts / 325 watts = 12.47 panels

Since there are no partial panels, you can round up this number. So, you need fourteen 360-watt panels to meet 75% of your energy needs. As you observed here, using the power output under NOCT provides more realistic results than when considering STC.

Method 2:

Once you know you have enough roofing space to mount the panels, identify the tilt angle and the direction the panels should be facing, and you can scan the code here to use the PV-Watts calculator to determine your monthly power output from the panels:

The PV-Watts calculator works as follows:

- Enter the address and click on the orange button on the right.

- On the open system info page, enter your previously-calculated DC size of the system.

- Pick a standard module.

- On array type, choose "fixed (roof)" if you have roof mounts or "fixed (open)" if you're dealing with ground mounts.

- Leave an allowance for system losses of around 15%.

- Enter the azimuth angle of 180 for southern-facing roofs along with the solar panels' tilt angle, which would almost always be your roof's tilt angle.

Now you can click the orange arrow on your right to obtain a monthly solar system output.

The PV-Watts Calculator provides an accurate breakdown of your energy output based on your location and the characteristics of your building.

Important Hints for Designing Your Solar System:

To get an accurate estimate for your solar system, you have to take into account the size of the panels that fit your solar design, the type of roof mount, and the direction to mount the panels. You can fine-tune your system design through:

1. Selecting Type of Panel Mount

Your location and surrounding objects are the most important contributing factors in determining your type of mount.

Your location: if you can't install the panels facing south within the preferred angle, then you have to add more panels to your system. Solar panels should always face the equator. If you live in the Southern Hemisphere, you should install the panels facing north. A sloping roof maximizes the efficiency of the solar panels.

Your surroundings: if you have tall surrounding objects or any potentially tall tree nearby, there is a high chance that your system's output will be less than estimated above. To overcome this problem, you should either remove the cause of potential shading or increase the number of panels. In these conditions, it is wise to consider micro-inverters and power optimizers in your design rather than conventional string inverters.

Roof mounts not optional: if mounting on the rooftop is not an option, you can try a ground mount or use pole-mounting solutions. If you want to lay them on the ground, you can position the panels in any direction. This helps to maximize the sun's exposure.

2. Roof Characteristics

Your roof orientation (azimuth angle) and pitch (tilt angle) determine the direction and angle of the solar panels. If your panels are installed facing east or west, they will produce an output that is less than that produced by panels facing south. In this case, you have to increase the number or efficiency of your solar panels in order to account for that.

Odd-shaped and small rooftops are important factors when determining the size of the solar system. If you have a large rooftop, you can purchase several large panels that generate the required energy output.

However, if you have a smaller rooftop, or most areas of the rooftop are partially shaded, then you must purchase a few high-efficiency, cost-effective panels to provide you with the target energy needs. Later, you can increase the number of panels to accommodate your increasing energy demands.

How to Size Hybrid Solar Systems

The hybrid solar system is a grid-tied system tied to a battery bank. So, it possesses characteristics of both grid-tied and off-grid systems. Since you are connected to the grid, there is no reason to worry much about your daily consumption. Additionally, in the event of an outage, you can take advantage of your backup.

In most hybrid systems, a subpanel, or essential load panel, is designed in such a way that you can supply your essential electrical devices. Below, you can see the steps needed to design a hybrid system:

Step 1: Estimating your Energy Needs (As Previously Explained)

By using the example below, we can better explain the different steps:

Assume you live in a house in California with 5.2 peak sun hours, where you consume 1100 kWh per month. Your system is supposed to be supported by a 12 V battery and 200 Watt Renogy solar panels with a power output of 170 under NOCT.

To determine the power consumption per day, we divide monthly usage by 30, or, if yearly consumption is available, divide by 365:

1100 kWh / 30 days = 36.66 kWh per day

36.66 kWh * 1000 = 36660 Wh

Step 2: Determine peak Sun Hours

Determine the peak sun hours based on your location using the previously-mentioned website. This will help you know how much energy the panels generate during peak sun hours: Wh per day/peak sun hours = required power output

36660 watt hour / 5.2 = 7050 watts

Number of panels needed =power output/module' rate 7050 Watt /170 Watt=41.47 (rounded up to 42) panels required to provide the whole daily consumption.

Depending on your essential loads and available roof space, you might just cover 60% to 70% of your daily consumption and install fewer panels than mentioned above.

Step3: Sizing Your Battery

Unlike off-grid systems, in hybrid power systems, you just need to cover your essential loads by the battery bank; therefore, you need to collect and list the consuming load of all your essential (desired) appliances to figure out how big your battery should be. Here is an exemplary list of most preferred appliances as the essential loads:

Loads	Power (Watts)	Duration (Hour)	Daily Energy Usage(Watthour)
(8) 10-watt LED lights	80	6	480
(1) 110-watt laptop	110	3	330
(1) 90-watt TV	90	3	270
(1) 300-watt fridge	300	12	3600
(1) 70-watt fan	70	2	140
(1)320-watt washing machine	320	1.5	480
			5300 Watt Hours

Since the battery cannot be 100% efficient, you have to account for an estimated battery loss of 15%. Divide watt-hours per day by 0.85 battery efficiency:

5300 Wh / 85% = 6235 Wh total battery capacity

Since batteries are rated in amp hours, not in watt-hours, convert the Wh to Ah by dividing the battery's capacity by the battery voltage (12 V, 24 V, or 48 V). As you're using a 12V battery, then:

Battery capacity = 6235 watt-hour / 12 V = 519.5 amp hour

Once you get the battery capacity, you can divide it by the battery's rating to know how many batteries you should use in your system. For instance, you need six 100-Amp Hour Lithium-Ion batteries that can store 600 amp hour (> 519 Ah), and consequently, you need an array that produces at least 7200 watt-hours (600 *12 volt = 7200 watt-hours). You should keep in mind that this amount of power is the minimum amount required to be supplied by the array. You can either go with this or scale it up to your entire daily consumption (36660 Watt-hours).

How to Size an Off-Grid PV System

Calculating your power needs for an off-grid system is different from a grid-tied system. Since you live off-grid, you have to focus on the daily power usage (kWh) instead of the monthly or annual electricity bills. An off-grid system makes you energy-independent since it can cover your day-to-day energy needs. For this reason, your system should offset 100% of your energy consumption demands and store more energy to run your home smoothly.

Without an electricity bill to guide you, you have to start by listing all the major appliances in your home and how much electricity you use on a daily basis. This will help you determine your load.

You can use the following load evaluation calculator to determine the size of your off-grid system:

Alternatively, you can check how much electricity each appliance consumes from the appliance electrical consumption guide or the energy guide sticker.

Let's look at the step-by-step procedure regarding how to size your system based on your location and your energy needs.

Step 1: Determine Your Energy Needs

You need to evaluate the energy needs for each piece of equipment you have in your home. Have a look at the user manual to confirm the device power consumption rate, or confirm the rate from the manufacturer's website.

Once you obtain the power consumption rate, you can multiply this figure by the number of hours you run the item every day.

For example, if we have a 300-watt fridge that runs for 14 hours a day, an oven rated at 2 kW (2000 W) that we intend to use it for half an hour each day, and we need two 10-watt light bulbs that run 24 hours per day, the total load list will be as follows:

Load (Watt)	Duration(Hour)	Power(Watt-Hour)
300-watt fridge	14	4200
2000-watt oven	0.5	1000
10-watt light bulbs * 2	24	480
		Total power: 5680

Step 2: Add Inverter Load

If you're using an inverter to convert DC to AC power for your consumption, you have to account for inverter and system efficiency losses. Inverters consume a fraction of generated power when running; therefore, you have to add the consumption rate of the inverter to your daily total. Different types of inverters have different consumption ranges. So, have a look at your inverter spec sheet to determine how much your inverter consumes. If your inverter consumes 30 watts and runs for eight hours in a day, for instance, you need to add that to your power load.

Inverter load = 40 W

Watts * 8 hours = 320 Wh

Total load = 5680 Wh + 320 Wh = 6000 Wh

You also have to account for inefficiencies in your system. Efficiency losses range from 5% to 15%, depending on the specific type of inverter and how much load is connected.

This is important when sizing the battery- hence the reason you should buy a quality, efficient inverter.

Step 3: Calculate Battery Size

Batteries store the collected solar energy for later use. The size of the battery depends on the required backup power to effectively run your appliances. During winter months, when there is not enough sunlight to power the panels, you can rely on a backup generator since battery storage might not be enough.

When choosing your battery size, you have to consider system inefficiencies and temperature coefficients associated with your off-grid system. These inefficiencies have an impact on solar output, and the rate of inefficiency depends on the solar equipment and the system design. You have to compensate for these inefficiencies by oversizing your solar panels appropriately. Typically, the battery's voltage output is 12 V, 24 V, 48 V, or 120 V; therefore, the first step is to select the battery voltage. Lower voltages, like 12 V, work perfectly for smaller systems, while 24 V and 48 V better suit a medium and large system, respectively. Another consideration is the inverter and charge controller capacity, solar array configuration, and wire sizes.

Assuming your total daily power consumption equals 6000 Wh, you then have to account for inverter inefficiencies. You can look at your inverter spec sheet to determine its efficiency. For example, if your inverter's efficiency rate is 90%, you will need to add 10% to your daily power consumption. So, the amount of energy drawn from the battery to run your load through the inverter will be:

6000 Wh * 1.1 inefficiency = 6600 watt-hours

Next, you have to account for temperature changes in your battery's capacity to deliver power. If you're using a lead-acid battery, you should expect a loss in battery capacity in lower temperatures. For example, if the battery temperature is around 20^0F during the winter months, you can multiply your battery capacity by 1.59. (Do not worry—this topic is covered in detail in chapter seven. Here, our main focus is on sizing the array.)

6600 Wh * 1.59 = 10494 watt-hours

You also have to factor in the efficiency loss that occurs when charging and discharging the battery. For lead-acid batteries, the efficiency loss is 20%, while lithium-ion batteries have an efficiency loss of 5%; therefore if you decide to add lead-acid battery to this system:

Minimum energy storage capacity of battery= 10494 * 1.2 = 12592.8 Wh

Days of Autonomy

After sizing your battery, you will also be able to know how much power you need to fully recharge it. You also have to record battery autonomy; that is, the number of days you want your battery to store energy. This is considered to compensate for consecutive cloudy days when the solar array receives significantly fewer peak sun hours. Generally, the days of autonomy should be between two to five days. If you're to run the battery for three days, you can multiply your battery capacity by the number of days of autonomy.

12592.8 Wh * 3 = 37778.4 Wh

As you can see, temperature changes and autonomy days significantly increase the size of your battery. And since batteries are measured in amp-hours (Ah) instead of watt-hours (Wh), you have to convert Wh to Ah. To do this, divide battery capacity by its voltage:

37778.4Wh / 12 V = 3148.2 Ah for 12 V battery bank

If we are supposed to use 100-watt, 12-volt lithium-ion batteries, then we need:

3148.2/100=31.5 rounded up to 32 batteries.

Therefore our solar array should be able to generate:

32 * 100 amp-hour * 12 volt = 38400 watt-hour

Further, when sizing a battery bank, you have to account for the amount of energy discharged from the battery. Lead-acid batteries have a discharge depth of around 50%.

This helps in extending the battery's life. Lithium batteries can allow deep discharge without affecting the battery life; therefore, if you use lead-acid batteries, you need to double the above number:

38400 watt-hour * 2 =76800 watthour

However, using lithium-ion technology will let you size your panels based on the former calculation (38400 Wh).

Step 4: Calculate Peak Sun Hours

Our next step is to calculate solar input based on where to install the panels (your location). You can scan the code below to identify your peak sun hours. During peak sun hours, the energy production is at its highest amount. Pay close attention to how much energy you consume each day. During December, for instance, power consumption tends to be very high, while in January, your consumption is lower. Alternatively, you can use the online resource map available for solar radiation from the National Renewable Energy Laboratory (NREL).

You should install solar panels in an area where they receive full sun exposure during the day and are free from any shading.

Step 5: Determine the number of Panels Required

Since you're able to determine your battery capacity, you can determine the size of your charging system. In this case, the charging system should store enough energy to replace energy drawn from the battery, and should also be able to account for system losses and inefficiencies.

Assuming the peak sun hours in your location is 5.2 hours and battery capacity of 76800 watt-hour as calculated above, then your array size will be:

76800 Watt hour/5.2 hours =14769.2 Watts

There is no system that is 100% efficient. So, you have to account for the system losses caused by inefficiencies in the system, such as voltage drop, rotation of the panels, and shading. Assuming the overall losses are 15%, the minimum PV array size will be:

14769.2 watts / 85% =17375.5 watts

By dividing the PV size by individual PV modules' power output, you can calculate how many panels are needed for your specific use. This PV array generates the amount of energy needed to charge your battery and run your loads for three autonomous days. Adding a backup generator will assist the solar array to keep your battery bank always fully charged in bad weather conditions and cloudy winter days.

Roof Sizing

So far, you already have an estimate on the number of panels you need to generate the required electricity needs. But the biggest question is, will all the panels fit on your roof?

Though you already know the square footage of your house, this doesn't mean the available roof space is fit for installing solar panels; however, if you have a big rectangle area facing south or west, that should be great for installing panels. The east-facing roof can also work in some cases.

Most rooftop designs, such as mustard roofs and gable roofs, do not have a simple rectangular shape. They have different geometry that interferes with the available mounting space. Sometimes, you may have enough mounting space, but it might be shaded by trees or nearby buildings.

Coming Up with the Design Layout

Most solar installers use professional software to come up with a design system layout for installing panels. Others use satellite images from Google Earth as the starting point. There are a number of websites that allow you to use online satellite maps to map out your roof area and determine square footage for installing the panels. One of the online tools you can use to estimate your roof area is the Map Developers.

By joining the site, you can try this area calculator to have a satellite view of your rooftop. You can follow the steps below:

Step 1: enter your address; you should enter your street address, city, and state.

Step 2: click on "zoom to address" to find the location of your house on the Google map.

Step 3: click "satellite"; the satellite view provides you with an actual view of the house rooftop.

Step 4: zoom in on your house by clicking the + icon on the lower right corner.

Having an image alone is not enough, however; you have to look for other factors such as shading and roofing issues on your actual house.

Step 5: map out your roof area by drawing some points on the rooftop. You can join the points to create a polygon. This is the area where you will mount the panels.

Step 6: get your square footage by calculating the areas around the polygon. The area is usually displayed at the top of the map.

Other Ways to Measure Your Rooftop Space

Another way to know the available space for panels is to climb up and take a direct measurement of all the open areas. This method can have obvious risks when walking on the rooftop, though.

For your safety, you can identify a rectangular section on your roof, then take measurements of the area from the ground. You only need to take measurements from the corners of the house where the roof's corners meet the ground. Measure the length (x) and width (y) of your house.

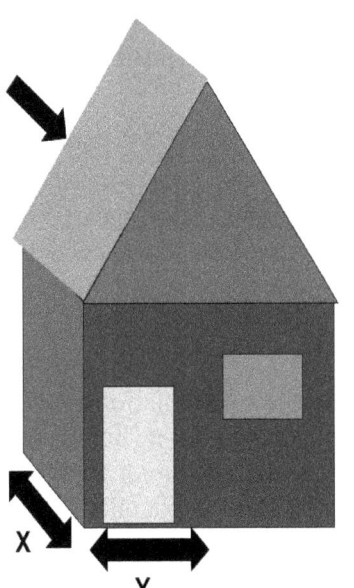

Multiply the x and y to get the area on the ground. If your x = 24 feet and y = 12 feet, then your mounting space from the ground will be 288 feet. So, how does this help you figure the number of panels to install?

Most PV modules measure 3 by 5 feet on average. So, for each panel, you need 15 square feet (3 * 5) space.

To know how many panels you can to install, divide the amount of square footage by 15.

288 square feet / 15 square feet = 19.2 (rounded to 19 panels)

You can install a maximum of 19 solar panels on your roof. As the number of panels must be an even figure for wiring and voltage purposes, 18 panels must be the accurate number. If your selected panel's size is different from the above, use the right size from the manufacturer's specification sheet.

Choosing Suitable Solar Panels

When buying solar panels, there are a number of factors you have to consider when looking for the best panels for your solar system, such as:

1. Cost

This is one of the major considerations for those who are on a tight budget. Panels come in the range of 200 to 400 watts, although there are other smaller panels on the market, especially if you're designing a small off-grid system.

The cost-per-watt helps you compare the prices of the panel with the level of energy output; that is, divide the price of the panels by the amount of output generated to know whether it is worth the price.

2. Efficiency

Efficiency ratings present the amount of sunlight that a panel converts into usable energy. Most panels have an efficiency rating of between 14% and 22%. The efficiency of panels directly influences the panel's output.

3. Temperature Coefficient

Temperature coefficients have an impact on the output of the panels. The coefficient is a measure of how much efficiency is lost for temperatures above or below the ideal testing conditions.

Manufacturers test panel output in a climate-controlled factory as well as real-world weather conditions. Sometimes, your system lags behind the efficiency level based on climatic conditions during that period.

For example, if your local highest ambient temperature is 65^0C, and the temperature coefficient is shown as 0.4% watt power in the product's datasheet, then you can multiply the coefficient by the temperature difference ($65-45=20^0C$). Now your panels are operating at 20 * 0.4% = 8% below the rated efficiency rate. Note that 45^0C refers to the ambient temperature under NOCT.

If you live in places that experience high temperatures, you should consider monocrystalline solar panels, which can perform better under high temperatures. Extremely hot areas can also reduce the panel's output because excess heat reduces the panel's efficiency.

4. Warranty

You should also consider the performance and workmanship warranty of the panels. The performance guarantees that the panels will work above the stated efficiency. Most manufacturers guarantee 80% efficiency for over 25 years, while the workmanship warranty covers defects and other physical problems with the panels.

A workmanship warranty is an indication of the reliability of the panels and is usually shorter than the performance guarantee. The standard workmanship warranty is usually 10 years, though some companies do offer a 20-year guarantee.

5. Company's Rputation

Another factor you have to consider is the longevity and reputation of the manufacturing company. Look for customer reviews about the company. If the company demonstrates a stable track record, you can rest assured you're getting value for your money.

Examples of Cost-Effective Solar Panels on the Market:

1. SunPower Maxeon 3

SunPower Maxeon 3 is a 400 W panel that ensures you get the best solar output in your home. The panels can deliver 35% more energy from the same space for over 25 years when compared to conventional solar panels. They have an expected lifespan of 40 years and the rate of return is less than 0.005%.

These panels have the lowest degradation rate, and they are four times stronger and more reliable than conventional panels. This makes them the market leader with an efficiency of 22.6%. The back-contact conductivity of Maxeon cells makes them absorb more sunlight since there are no metal gridlines. As a result, they generate more energy than other types of panels.

2. SunPower SPR-X22-360

The X-series is another efficient and sought-after panel in the US because of its unmatched power and performance. It has an efficiency of 22.2%. The panel is ideal in situations where there is premium space, and there may be future expansion of the panels.

It provides maximum performance given real-world conditions, especially in higher temperatures. The panels record a 0.25% degradation rate each year, making them more reliable.

They come with an integrated micro-inverter that converts the DC energy to AC and also optimizes the performance of each PV module on the roof.

3. Panasonic VBHN 330SA17

Panasonic PV modules feature an innovative monocrystalline cell structure that consists of 96 cells and produces 36% more energy when compared to 60-cell panels. The 330 W PV modules produce an extra 10% of electricity when the temperature increases.

The PV module efficiency is 21.76% while cell efficiency is 19.7%. Manufacturers of these panels guarantee a 100% on power rating, a performance guarantee for 20 years, and a 10-year workmanship warranty. If you're on a budget, you should consider this type of module.

Where to Mount Solar Panels

Once you buy your solar panels, it's time to decide where to mount them. Your decision on the type of mount to use depends on your installation space, budget, and energy needs. There are four different types of solar mounts:

Rooftop Mount: this mount is the most common type of installation that allows you to mount the panels in an existing roof structure. It is also the less expensive type of mount. If you install panels on the rooftop, there are only limited opportunities for expanding your array size in the future, unlike a ground mount, which doesn't have such restrictions.

Ground Mount: ground mount allows you to fix a foundation on the ground level. Though it occupies a lot of space, ground mounts are easy to install and provide you with maximum control over the orientation of the array of panels to maximize energy production.

A ground mount system allows you to align the panels at an optimal angle so they face directly toward the sun.

This makes them more efficient as they have maximum exposure to sunlight. In addition, the panels are raised off the ground, and this allows airflow under the panels to cool them. As a result, the panels generate more energy.

Carport Mount: these panels are built to cover a parking area and provide you with more efficient use of your living space than the ground mount panels.

Pole Mount: this allows you to mount an array of panels on a single pole that is elevated higher off the ground.

Pole mounts may incorporate a tracking system that automatically tilts the panels to the direction where they can receive optimal sunlight exposure.

The rooftop and ground mounts are the most common racking systems, and each has various merits based on your solar project specifications.

No matter the type of mount you choose, make sure the panels get maximum sun exposure. For example, if you live in the US, where the sun leans to the south, installing the panels facing true south will produce the best results. But if you live in South America, installing panels facing true north will produce better results.

Pros and Cons of Roof Mount PV System

Pros:

- Less expensive
- Requires few installation tools
- Requires lower labor costs
- Easy to utilize unused roof space

Cons:

- Hard to troubleshoot panels and other accessories
- High temperatures reduce panel output
- Hard to access the roof, especially slippery or steep roofs
- Available roof space constraints limit the size of the panels
- Can cause water leakage via the installation holes

Pros and Cons of Ground Mount PV System

Pros:

- Easy to install and access
- Easy to clean
- Easy to troubleshoot panels and other accessories
- More efficient
- Easy to expand your system
- Strong racking system
- System not defined by the roof dimensions

Cons:

- Expensive
- More labor-intensive
- Requires more installation tools
- Requires more installation space

Your decision on whether to go for ground or rooftop mount depends on:

• Budget: how much are you willing to spend on designing the system, including upfront costs? A rooftop mount is ideal if you're looking forward to maximizing your return on investment (ROI).

• Type of soil: if the soil in the area makes it difficult to dig, then you can consider a rooftop mount.

• Future expansion: do you plan to expand your solar system? If you want to go off-grid, a ground mount one will be more suitable since it provides room for expansion.

Mounting Systems

There are different types of mounting systems that secure solar panels on your rooftop or ground. If you're new to the solar installation, identifying these systems help you install your panels easily. They include:

1. Sloped Roof Mounting System

Installation of solar panels in most homes is usually on sloped rooftops. There are various options for mounting the panels in this kind of roof; railed, shared rail, and rail-less. All these options require certain penetration or anchoring to securely attach the panels to the rafters or directly to decking.

Rail System

Most of the residential solar systems use rails mounted on the roof to support several rows of panels.

Rows of panels are usually placed in a portrait position; rails and clamps are used to connect them. Rails are secured by using screws or bolts.

Rail-Less System

In a rail-less mount system, the panels are connected directly using screws and bolts that penetrate through the roof instead of attaching the panels to the rails.

This reduces the manufacturing and shipping costs of the system and makes installation much faster.

The panels require the same number of attachments penetrating into the roof, just as in the rail system.

If you're using rail-free mounting, you can position the panels in any orientation. This type of installation is more suitable for roofs with irregularities where installing rails will result in fewer panels to be installed.

Shared-Rail System

On the other hand, the shared-rail mount system works the same as the rail system, but has a different number of rails. In the rail system, every two rows of panels connect to four rails, while in shared-rail, the middle rail is shared between two rows. So ,only three rails are needed. In this case, you connect two rails on the edges of the panel and clamp the other rail in the middle of the two rows of panels.

You can use any orientation to position the rails, and consequently, the panels. After determining the accurate position of the rails, installing panels is quick and straightforward since you don't need mid and end clamps, or more rails as is used in a regular railed system.

2. Flat-Roof Mounting System

These are mostly used in building structures that have flat rooftops, especially commercial or industrial buildings. One of the advantages of a flat-roof mounting system is usually the large space available for installing the panels. This makes it easy to install the panels.

This system uses a flat roof, ballasted mounting system approach. It uses "foot" as the base assembly for the mounting structure. The foot is tray-like equipment positioned on the rooftop with a tilted design. It holds ballast blocks at the bottom and PV modules at the top and bottom edges. The panels are tilted to an angle that captures the most sunlight based on the location.

The roof load limit determines the amount of ballast needed to install the panels. Clamps or clips are utilized to fix the panels securely on the mounting system. If you have a large, flat roof space, install the panels facing south. You can also generate electricity with panels facing east or west. Furthure, you can use a dual-tilt configuration to utilize your roof space.

3. Solar Shingles System

As more people are now interested in a unique and aesthetic solar installation design, the shingles mounting system is increasing in popularity.

Solar shingle is a building-integrated PV (BIPV) system in which the solar panels are part of the built-in structure.

Tesla Solar Roof does not require a separate mounting system because the panels are already integrated into the roof (they're part of your roofing structure).

Roof Mounting Installation Process

Now that you already know the different mounting systems, let's look at how you can carry out the installation of the panels.

In this instance, we will discuss the steps to follow when building a railed mounting system.

Step1: Prepare System Components

Start by assembling the required components for installing the panels: drilling tools, rails, bolts, clamps, screws, fall protection kit, pencils, etc.

Step 2: Check the Guidelines to Know Where to Position Panels

Take into account the installation limits related to where to position rails and panels from the edges of the roof in your solar panel's guideline.

DIY SOLAR POWER FOR BEGINNERS

Step 3: Determine where to Drill Holes on the Roof

Measure the distance between the pre-drilled holes on your solar panels to match with the distance between the rails on the roof.

Step 4: Find Rafters on the Rooftop

Rails must be attached to the roof rafters by screws; therefore, we need to find the rafters. Drill a spot hole into the attic through the roof. Find this drill position from the inside, and then adjust the truss at the center in order to fix the rails in position and provide base support.

Drill your first hole to confirm the exact center of the rafter, then mark another spot on the same rafter by using the same distance between the first confirmed hole and the edge of the roof (X). Connect the two spots on the rafter using a chalk line. Depending on the number of rows and railing system, you may need to add one or two more holes on the same rafter. The distance between the first two holes on the first rafter is determined by pre-drilled holes of your panels.

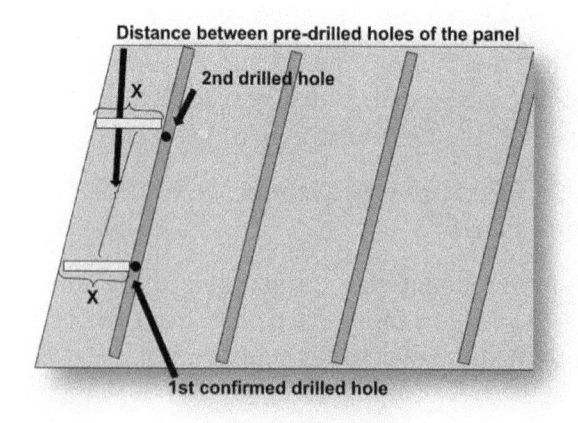

Once you chalk your first rafter, you can easily find other rafters by measuring the standard 24-inch distance between rafters. Chalk as many rafters as you need, then drill the holes on the rafters using the bit that fits your flashing's lag bolts.

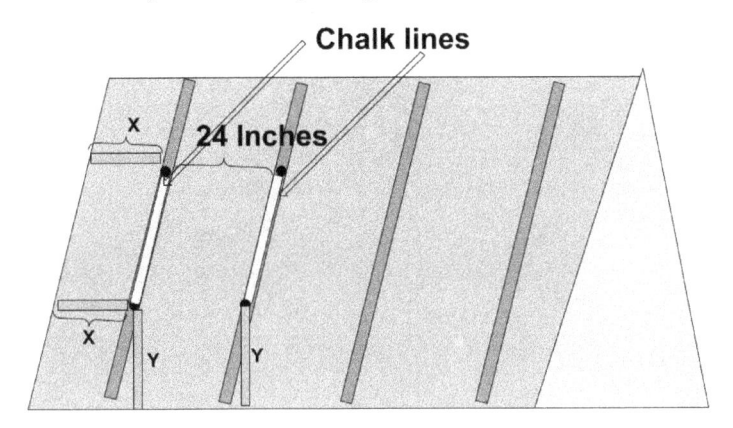

Step 5: Mount the Flashings

After drilling the holes in place, mount the flashings that act as a supporting structure to aid in the installation of the rails on the roof. Using a ratchet, fasten the lag bolts to tighten the flashings in place. Most systems have a cap that sits on the bolt to seal as well as prepare a place to keep the rails.

Step 6: Mount Solar Racking System

In this step, put the rails in place and properly line up and elevate the rails using the flashings. Rails are kept in place by their brackets connected to the flashing caps in some railing systems. In other systems, rails just click into the flashing's connector, making the system installation much easier.

Rail   Flashing

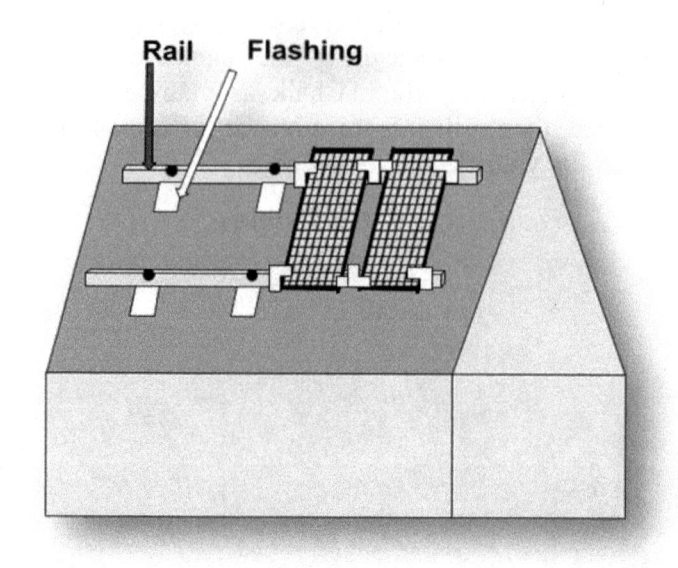

Step 7: Wiring Panels

Attach the wires to the management clips and fasten the grounding bolts attached to the copper wire connected to the grounding system of the house.

Step 8: Install Micro-Inverters or Power Optimizers

You can connect each panel to a micro-inverter to boost the efficiency of the panels. Secure the wires using the management clips.

Step 9: Secure Panels to the Mounting System

The last step is to tighten the clamps and T-bolts to ensure the panels are securely fixed on the mounting system. There are two types of clamps for securing panels on the rails: mid clamps and end clamps. They both have a T-bolt structure that goes in the rails and turns 90 degrees to get locked. Mid clamps are located between two panels, while end clamps keep the panels at the two ends of the rails.

Chapter Summary

Learning how to read panel specifications helps avoid confusion and lets you understand the normal operating conditions of the PV modules. Using the current-voltage (I-V) curve, you can easily understand the performance information regarding efficiency, maximum-rated power, maximum voltage, and maximum current, among other factors.

Furthermore, the datasheet provides information on how to install and operate the panels. You're also able to size your panels based on your daily load. Sizing solar panels is an important step to achieving energy independence. Once you estimate your energy needs, you can calculate the required PV output and the number of panels you need to meet your energy demands. The performance of the PV modules depends on tilt angle, sun peak hours, the orientation of the panels, and shading effects.

Next, you have to estimate the design of your solar system. This is highly dependent on your roof type, roof characteristics, choosing the right type of solar panels, and calculating solar power output. Determine where to mount the panels.

If you're planning on rooftop mounts, you have to size your roof to make sure you have enough space to install the panels. Once you take into account all the factors, you can go ahead and install the panels.

In the next chapter, you will figure out how batteries can help you get more energy-independent.

CHAPTER SIX: Solar Charge Controllers

Charge controllers are an essential component of battery-backed PV systems. In fact, off-grid systems, as well as DC-coupled hybrid PV systems, should utilize a controller in order to regulate the process of charging the batteries.

The DC power generated by your solar array flows to the charge controller to charge the batteries.

This chapter will walk you through the most essential features of charge controllers and guide you on how to select, size, install, and program your device properly in order to match your PV system.

What Is a Solar Charge Controller?

Basically, a solar charge controller's main task is to regulate voltage and current from the solar array in order to protect batteries from overcharging, as well as over-discharging.

Charge controllers are critical components used in off-grid and DC-coupled hybrid PV systems to regulate the PV output for battery charging.

As explained before, grid-tied and AC-coupled hybrid PV systems do not require a charge controller. And that is because no battery is used in the former one, and an inverter/charger regulates the battery charging in the latter one.

Why Do We Need a Charge Controller?

Charge controllers carry numerous functions in PV systems; however, they play two critical roles in relation to batteries:

As explained in the previous chapter, a solar array with a nominal voltage of 24, for instance, is often supposed to store its excess power in a matching nominal (24-volt) battery; however, this setup will generate a Vmpp (maximum power point voltage) of around 36 V, which will definitely damage the battery bank if directly connected to it. Here is where the solar charge controllers come in handy. Most 24-volt batteries need a voltage of around 27 to 28 volts (larger than their nominal voltage) to get fully charged. The same rule applies to 12- and 48-volt batteries.

Additionally, at nighttime, when solar panels generate no power and therefore no voltage, the higher voltage of batteries when compared to the array, can drain the batteries and push the current toward solar panels. Charge controllers prevent this potential reverse current that can harm your batteries. Other than the main tasks mentioned above, most charge controllers provide a range of functions, such as load control and lighting. Below are some of the optional features you may find helpful in some controllers:

- *LED or LCD displays:* are available in some controllers. Some controllers like Victron can also be monitored by connecting to your phone.

- An alternative to a display is an *MT 50* screen which is included in some models and can be connected to some other models, such **as the Epever charge controller**, for programming the device, as well as displaying the charging data.

- *Temperature compensation*: most modern controllers can adjust the battery input according to the battery's temperature and temperature correction voltage to improve the battery-charging process.

- *Low voltage disconnect:* when your DC loads are connected to these charge controllers, this feature will protect the battery by disconnecting the current while the voltage is too low.

- *Lighting:* some controllers can turn the lights on and off at dawn and dusk, respectively.

Charge Controllers and Stages of Charging

Charge controllers help batteries undergo the stages of charging properly. There are three to four stages of charging, depending on the type of battery.

- *Stage 1, bulk charging.* When the sun is out, the charge controllers will send all the available current to charge the battery.

- *Stage 2, absorption.* As the battery reaches the regulation voltage, the controller keeps the voltage constant. The battery is therefore protected from overheating and over gassing (in lead-acid batteries).

- *Stage 3, equalization.* Only flooded lead-acid batteries undergo this stage. The periodic boost charge in this stage can lead to better stirring of the electrolytes and completing the reactions, as well as leveling the voltage of the battery.

- *Stage 4,* float *charge.* This is the stage where the battery is fully charged, so the controller reduces the voltage to avoid over-gassing and overheating.

The graph below illustrates the above-mentioned stages of charge:

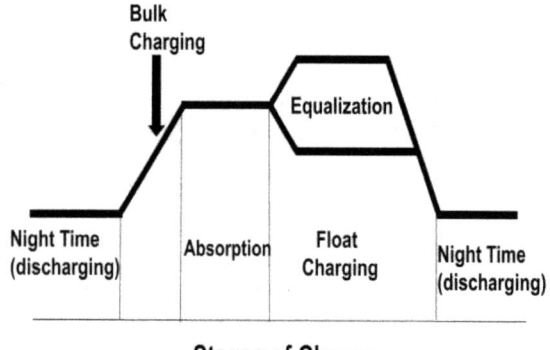

Stages of Charge

Charge Controller Technologies

Charge controllers are available in two different technologies, PWM and MPPT.

PWM Solar Charge Controllers

A Pulse Width Modulation or PWM solar controller directly connects the solar array to the battery. During bulk charging of the batteries in the daytime, the continuous connection between the array and the battery will pull the array's voltage to the battery's voltage. As the battery is getting charged, its voltage will rise, and so does the array's voltage. That is the reason you can only use a matching nominal voltage of battery and array when using a PWM charge controller.

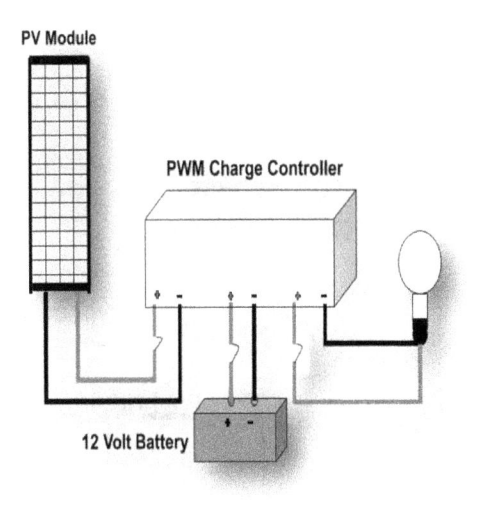

To explain the essence of matching nominal voltage when using a PWM charge controller, we can connect a PWM controller to two matching and non-matching PV systems and observe the results.

But first we need to identify the Voc and Vmp of the two PV systems. The table below shows the approximate open circuit voltage and maximum power voltage of three different panel types. Manufacturers include the exact figures in the datasheet for each nominal voltage of PV modules.

Nominal Voltage	Number of Cells	Open Circuit Voltage (Voc)	Maximum Power Voltage (Vmp)
12	36	22	18
20	60	38	30
24	72	44	36

Now we can compare a 12-volt and a 24-volt module charging a 12-volt battery using a PWM controller:

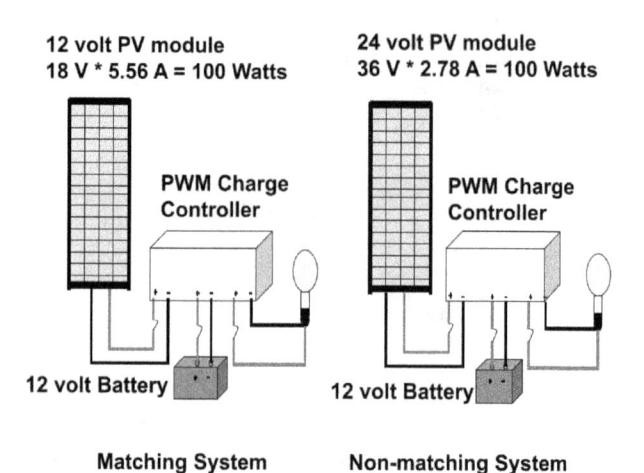

12 volt PV module
18 V * 5.56 A = 100 Watts

24 volt PV module
36 V * 2.78 A = 100 Watts

PWM Charge Controller

PWM Charge Controller

12 volt Battery

12 volt Battery

Matching System

Non-matching System

- **100-watt, 12-volt solar panel to 12-volt battery**:

As you can observe in the left-hand diagram, the 12-volt battery output power will be as calculated below:

P = I * V

P = 5.56 amps * 13 Volts = 72.28 watts

Therefore, the system's efficiency would be:

72.28 watts / 100 watts = 72.28 %

The efficiency rate of PWM controllers is less than the premium (80-90%) efficiency of MPPT controllers, which is due to the direct connection and not benefiting from the maximum power tracking in MPPPT controllers. **Note that** a 12-volt nominal battery would have a voltage around 13–14 V when fully charged. And that is the reason we used 13 V as the battery's voltage here.

- **100-watt, 24-volt solar panel to a 12-volt battery**:

In the right-hand diagram, the 12-volt battery output would be as follows:

P = 2.78 A * 13 V = 36.14 watts

The system efficiency would be 36.14%. As simply calculated above, the PWM controller only performs fairly well when the PV module and battery nominal voltage are the same. This example is intended to explain the way PWM controllers work. They should always be connected to matching voltages of the solar array and the battery.

As explained, even when PWM controllers are connected to matching panels and batteries, they perform at an efficiency rate of around 75%, which would be lower than the decent MPPT controllers.

PWM controllers are the least expensive controllers on the market. Some PWM controllers are available in multiple nominal voltages, meaning that they can charge a 12-volt battery using a 12-volt array and a 24-volt battery storage using a 24-volt nominal array. Note that PWM shouldn't be used with a 24-volt array to charge a 12-volt battery because it will result in energy loss.

How to Size a PWM Controller

Sizing a PWM charge controller is pretty straightforward. They are rated in amps as well as the previously-mentioned nominal voltage. By looking into your solar panel specification sheet, you can find the short circuit current (Isc) of the panels. By multiplying the Isc of the panel by the number of parallel strings, you can find the maximum total current of your solar array.

According to the National Electrical Code, you should also consider two safety factors.

> 1. 25% for over irradiance conditions when sun exposure is more than Standard Test Condition (STC).
>
> 2. 25% for more than three hours of continuous use. This is often the case with PV systems.

Total amperage = number of strings in parallel * panel's Isc * 1.25 (over irradiance) * 1.25 (continuous use).

To better explain the process of sizing, let's try an example:

Assume that we have two strings each composed of **two Renogy solar panels** rated at 100 watts and 12 volts, to set up a 24-volt nominal solar array. The panel's specification sheet indicates an Isc of 5.86 amp and 22.3 volt Voc. The maximum current is calculated as below:

2 strings * 5.86 amps (Isc) * 1.25 * 1.25 = 18.28 amps

So, a **20-amp Renogy charge controller** can easily handle our off-grid setup. Do not forget to check if the selected controller can handle our PV system's voltage.

MPPT Charge Controllers

As their name implies, Maximum Power Point Tracker or MPPT controllers, have the capability to track the Vmp (maximum power voltage) of the solar array to regulate it into a usable voltage to charge the battery. By using a MPPT, you can have a 20- or 24-volt nominal solar array charge a 12-volt battery bank.

These controllers are more common and also more expensive; however, they can increase the system's efficiency by 20% to 30%. As explained under Kirchhoff's law, the power that enters the MPPT equals the power that comes out of it; therefore, when charging a 12-volt battery via a 24-volt nominal voltage array, the MPPT reduces the voltage to around 14–15 volts and increases the current so that the power output remains equal:

Power (Watts) = I (Amps) * V (Volts)

MPPT controllers can let you be more flexible while choosing your system's voltage. For instance, you can use a 20-volt nominal voltage array to charge a 12-volt battery, or two of those panels connected in series to charge a 24-volt battery, or even three 20-volt panels in series to charge a 48-volt battery.

The most significant advantage of increasing your array's voltage is the fact that higher voltages mean lower current, hence the smaller wire size and lower cost. Additionally, you need lower battery voltage, and this can significantly decreases your upfront costs as well.

Size an MPPT Charge Controller

To select an MPPT controller, you need to consider three separate features of your PV system including current (amps), voltage, as well as the temperature.

1. Size the Amp Rating

MPPT charge controllers often reduce the array's voltage to the charging voltage of the battery. Since the power into the MPPT is equal to power out of it, this will result in a rise in the output current.

Therefore, by dividing the array's power by the battery's voltage, we can estimate the amperage.

Don't forget to multiply the number by 1.25 (NEC safety factor).

So:

Amp rating = total power output/battery voltage * 1.25 (NEC safety factor)

2. Size the Voltage

Although the MPPT controller can handle higher nominal voltages than the batteries, the Voc (short circuit voltage) is the limiting factor. **For instance,** a 24-volt solar panel has a Voc of around 44 volts; therefore, we need to select a controller that can handle the open circuit voltage (Voc) of the string. As a reminder, the Voc of the string is calculated as below:

Voc of the string = Voc of panel * number of panels in series.

3. Temperature Compensation

When exposed to low temperatures, solar panels generate voltages higher than the open circuit voltage (Voc) measured under Standard Test Condition (STC).

Lower temperatures can increase the Voc of each string of panels by approximately 0.03% per each degree Celsius below 25°C. This feature is referred to as the Voc temperature coefficient of the solar panel, usually mentioned in the specification sheet.

For instance, if a solar panel has a Voc temperature coefficient of -0.03% per degree Celsius, it means that we need to add 0.030% to the Voc for each degree Celsius below 25°C, and decrease the Voc by 0.030% for each degree Celsius above 25°C.

Higher temperatures than 25°C have no effect on sizing the controller; however, it must be considered when programming the controller.

An alternative way to find the highest Voc on cold winter days is to check the NEC, table 690.7 A, to find the temperature correction factor for each temperature range below 25°C. For instance, for ambient temperature range of - 6°C to -10°C the factor is 1.16. Simply multiply your array's Voc by this number to calculate the highest Voc possible.

We can formulate the sizing procedure as below:

Maximum volts = number of panels in series * Voc of the solar panel + [solar panel's Voc temperature coefficient * (difference between 25°C and the lowest ambient temperature)]

The example below better explains how to size MPPT for your off-grid setup:

Assume that we want to connect two **12-volt, 190-watt HQST solar panels** with Voc of 24.3 volts and Voc temperature coefficient of $-0.33\%/^0C$ in parallel to a 12-volt battery bank. As explained above:

MPPT amp rating = total power output / battery voltage * 1.25 (NEC safety factor)

So:

MPPT amp rating = 2 * 190 watt / 12 volts * 1.25 = 39.58 amps

Therefore, the controller should at least handle 39.58 amps. If the lowest ambient temperature in the area is -10°C, then,

Maximum voltage = number of panels in series * Voc of solar panel + [solar panel's Voc temperature coefficient * (difference between 25°C and the lowest ambient temperature)]

So:

MPPT Volt rating = 2 * 24.3 volts + [0.33% * (25°C-(-10C))] = 48.6 + 11.55 % = 54.21 volts.

In this case, a **60-amp Renogy MPPT** with its maximum voltage input of 150 volts, can easily regulate the charging of batteries.

Choose the Right Solar Controller: PWM versus MPPT

The PWM is an inexpensive option for smaller systems and is perfect for PV modules with nominal voltage of 12, 24, and 48 volts to charge 12, 24, and 48 volt battery banks, respectively.

The MPPT controller, on the other hand, is more suitable for larger systems where additional percentages of efficiency will lead to generating tens or hundreds of more watts per hour. Additionally, only MPPT charge controllers can efficiently charge a 12-volt battery bank; for instance, using a residential 20-volt nominal PV module.

Other than considering the battery voltage, some important considerations must be taken into account to select the controller. The most important one is your battery type.

Most flooded and sealed lead-acid batteries are easily handled by most charge controllers; however, if you are charging a custom (DIY) lithium-ion battery, you will need a charge controller like **the Epever charge controller, which can be programmed using an MT 50 screen connected to it.**

MT 50 screen is a useful monitor with the ability to be connected to the controller via a cable. This will enable you to monitor and program your controller. Charge controllers, like the Victron MPPT, make the programming of lithium batteries even easier. You can connect your phone to the MPPT via Bluetooth and adjust the features accordingly.

Other available solar charge controllers that do not come with an MT 50 screen or Bluetooth connection options are not recommended to be used with custom batteries because they are hard to program accordingly.

Programming Solar Charge Controllers

Different battery manufacturers recommend various voltages for their products. Added to that, batteries may need compensation as the temperature goes higher or lower than 25°C; therefore, the charge controller must be programmed accordingly. You need to get familiar with the different features of controllers and batteries to program. These features are as follows:

1. Charge profile: the different stages of charging each battery undergoes is referred to as the charge profile and include: bulk, absorption, equalization, and float charging. Batteries need different voltages depending on their charging profile.

By looking inside your battery specification sheet, you can easily find the appropriate charging voltage for each stage.

2. Temperature compensation voltage: the chemical nature of lead-acid batteries indicates less activity in cold weather, and higher chemical reactions when it is hot; therefore, the input voltage of lead-acid batteries needs to be increased and decreased in lower and higher temperatures than 25°C accordingly.

The temperature compensation voltage is referred to as the amount of voltage (in volts) needed to be adjusted for charging voltage of lead-acid batteries in temperatures higher or lower than 25°C. This is usually mentioned in the battery's specification sheet. For most lead-acid batteries, it is typically around -3 millivolt per ^{0}C per cell. This means that a 2-volt (1 cell) battery in 24°C needs 3 millivolts more than the recommended voltage of the manufacturer to be properly charged.

So, first you need to divide the battery's voltage by two to see how many cells your battery contains. For instance, a 12-volt battery has 12/2 = 6 cells. In our case, and under 24°C, the 12-volt battery needs (12 / 2 volts * 3 millivolts =) 18 millivolts more than recommended to be charged in 24°C.

As emphasized above, only lead-acid batteries need the voltage to be compensated in ambient temperatures higher and lower than 25°C. The example below explains the process of adjusting the battery's charging voltage:

Assume you are using a lead-acid battery in an area where, in summertime, the battery experiences a high of 45°C; so, you need to program your controller as follows:

From the lead-acid battery specification sheet, we find

that the recommended float charging voltage should be 13.3 volts, while the temperature compensation voltage is -20mV/°C/cell.

It means that for each degree Celsius above 25°C, -20 millivolts should be added to the voltage of each cell, so:

45 - 25 = 20°C temperature difference.

20°C * -20 millivolts = -400 millivolt = -0.4 volts per battery cell.

Each battery cell equals 2 volts, so a 12-volt battery has 12 / 2 = 6 cells.

-0.4 volts * 6 cells = -2.4 volts should be added to the recommended charging voltage;

13.3 volts + (-2.4) = 10.9 volts.

Therefore, you need to program the controller to charge your battery with this voltage in summer time.

Similarly, in colder climates, the voltage should be compensated. In fact, the voltage will be increased to improve the chemical reactions of lead-acid batteries.

Important Hint:

Do not confuse the battery's temperature compensation voltage with the solar panel's Voc temperature coefficient.

The temperature compensation voltage consideration for lead-acid batteries is due to their chemical nature, making them more active in hotter climates, while less active in colder weather conditions, whereas the Voc temperature coefficient refers to the solar panel's Voc changes in different climatic circumstances.

> 3. ***Charging Rate:*** we need to set the controller in order not to charge our battery at a faster rate than it can handle.
>
> 4. For instance, a 100-amp-hour battery should not be charged faster than 100/2 = 50 amps, meaning that if you use a controller bigger than 50 amp, you need to change the charging rate.

Smaller controllers do not need to be programmed for charging rates.

Each individual battery has its specific charging voltages.

For instance, the following charging specification can be extracted from **a 12-volt 100-Amp-hour Battleborn battery.**

Absorption Voltage: 14.2 V to 14.6 V

Float Voltage: 13.4 V to 13.8 V

Equalization Voltage: 14.4 V (if applicable)

Absorption Time: 30 minutes per 100Ah battery bank

Connect the Charge Controller

Once you have learned enough about the different features and functions of solar controllers and selected the one that fits your solar setup, it is now time to connect it.

Charge Controller Installation Tools:

- Wire stripper
- Wire crimping tool
- Heat shrink gun
- Appropriate screwdriver
- Impact driver
- Ratchet
- Ammeter
- Wattmeter

First of all, connect the ground wire from your controller to the ground terminal of your load center. Using a wire stripper and crimper, connect the negative, then the positive terminals to the corresponding terminals of the battery.

Take safety measures, especially if you are working with a 48-volt battery. Do not forget that DC voltages of 40 or greater can cause death if they come into contact with your body.

Turn off the DC disconnect switch before connecting the negative, then the positive terminals of the array to the controller. Use an ammeter and wattmeter to check the current of the battery and the wattage of the array. If they are within your nominal range, you can turn on the DC disconnect switch. You are good to go now.

Chapter Summary

The solar controller's function is crucial to your battery health. If selected and adjusted properly, the controller will secure a long battery lifespan, appropriate charging and discharging and smoothly running PV system. In the next chapter, we will explain battery essentials.

DIY SOLAR POWER FOR BEGINNERS

CHAPTER SEVEN: Solar Battery Bank

Off-grid solar power systems, also known as "stand-alone" systems, are designed to operate independently of the grid. Since your solar array does not generate electricity during the night hours, for continuous power flow, the system needs to store some extra energy in the batteries.

Solar batteries carry out their task in PV systems by storing the power generated by the solar array. While there are some recent battery technologies with built-in inverters offering integrated energy conversion, in the majority of the existing systems, power is sent from batteries to inverter(s) to be converted to AC power suitable for AC appliances.

In daylight hours with high sun intensity (usually 10 AM to 2 PM), when your array's energy output is exceeding your consumption needs, batteries become helpful by storing the excess energy. The higher the battery capacity, the more power can be stored.

What Is a Deep-Cycle Battery?

A deep-cycle battery, like all the solar options, is designed to provide steady output for a significant period of time until discharging to its recommended limit. This type of battery is different from car batteries, which should always be ready to generate a burst of high-power output to start the car engine, for instance. As this definition implies, off-grid and hybrid PV systems can just utilize deep cycle ones. Different types of deep cycle battery technologies can be paired with solar systems such as lead-acid, lithium-ion, nickel-based, and flow batteries.

Types of Batteries

Lithium-ion and lead-acid batteries are the two most popular chemistries available for off-grid systems. They are, indeed, not only different in chemistry, but also in many other aspects, such as the cost, lifespan, and capacity. Lithium-ion technology is newer and superior to lead-acid in almost all areas apart from the price. While they offer higher efficiency, capacity, depth of discharge, and lifespan, they do, unfortunately, cost a lot more than their old-fashioned competitors.

Nickel-cadmium and flow batteries are other solar battery types. They're seldom used for residential or recreational purposes, and they won't be covered here.

Lead-Acid Deep-Cycle Batteries

This technology has been utilized for storing energy for more than a century. Due to their reliability and low cost, these batteries are still extremely suitable choices for small and medium-sized PV systems.

The main downside of these batteries is their maintenance, especially in flooded types, which needs ventilation, as well as frequent refilling. Ventilation considerations will lead to installation issues as well. Their shorter lifespan (seven to ten years) and depth of discharge are other disadvantages. These batteries should only be discharged up to 50% of their capacity to prevent from being **permanently damaged. Deka or Crown Battery may** be among your best options to select as a lead-acid battery.

Mechanics of Lead-Acid Batteries

A lead-acid battery consists of the following compartments:

1. Lead plates

2. Diluted sulfuric acid solution, known as electrolyte

3. Negative electrode (anode)

4. Positive electrode (cathode)

The lead plates work as a separator blocking a direct link between two opposite electrodes. During the period of charging, lead oxide is produced at the cathode, while during discharging, sulfate ions leave the electrolyte, and water is produced. If the battery is overcharged, the inflammable gas, hydrogen, may be formed. This increases the risk of an explosion, hence the importance of proper monitoring and ventilating the batteries.

Types of Lead-Acid Batteries

1. Flooded batteries are the traditional type used for starting engines in motorcycles, golf carts, as well as deep cycle solar power systems. As the battery dries up, the user can open the lid to add distilled water.

2. Sealed batteries utilize the very same mechanics as flooded ones; however, a sufficient amount of liquid has been added before sealing the battery. As the name implies, you cannot add distilled water to this type.

3. Valve regulated lead-acid (VRLA) batteries are sealed lead-acid batteries equipped with a regulating mechanism to let oxygen and hydrogen gasses escape safely from the battery.

4. Absorbed gel matt (AGM) is a variant of VRLA lead-acid batteries, constructed with a more advanced technology for keeping electrolytes suspended around active plates in order to provide a decent depth of charge and discharge. Solar and storage purposes are among the uses of this type.

5. Gel lead-acid batteries utilize the same electrolyte suspension mechanism, although due to the addition of silica to the ingredients, they are not considered a wet cell. While they offer a slightly longer lifespans in hotter weather conditions, there is more chance that they'll have a shorter lifespan if charged with the wrong voltage.

Lead-Acid Battery Features and Terminologies

To get our feet wet with the topic, let's dive into some terminologies first. You may need to use some of them for proper use and maintenance of your PV system battery bank.

Total capacity is defined as battery's maximum energy storage and measured in amp hours. A 100-amp hour battery provides half the capacity of a 200-amp hour battery.

Specific gravity is defined as the ratio of electrolyte density when compared to the density of pure water.

Internal resistance is known as the resistance caused by the battery itself.

Cranking amps (CA) is defined as the current generated by the battery in 32°F for 30 seconds with the voltage being 7.2 volts.

Cold cranking amps (CCA) is defined as the current generated by the battery in 0°F for 30 seconds while voltage is 7.2 volts.

Reserve capacity (RC) explains the amount of time in minutes during which a battery can discharge 25 amps at 80°F while still remaining above 10.5 volts.

Depth of discharge is defined as the percentage of battery storage being safely utilized before harming the battery.

Lead-acid batteries will get seriously damaged if drained more than 50%; in fact, the ideal depth of discharge for lead-acid batteries is around 40%.

This will secure a long lifespan; however, you will need larger battery storage to address the total electricity needs of your appliances.

Lithium-ion batteries however can be discharged up to their whole capacity.

C rate (load factor) is defined as the rate at which the battery discharges its total capacity at 25°C, during a certain period of time (5, 10, or most often, 20 hours). A 100-amp hour, 20-hour rated battery, for instance, will have a discharge current as follows:

Discharge current (amps) =100/20 = 5 amps

Discharge current here (5 amps) indicates the current at which the battery power is emptied in 20 hours. Now, if we divide the current of discharge by the batteries total amp-hour, we will find the C rate:

C rate = 5 amp / 100 amp-hour = 0.05°C rate (load factor)

Nominal voltage of the most popular batteries available on the market is 12, 24 or 48. The voltage of a battery is related to the voltage of other PV components and does not determine the storage capacity of the system.

Lead-Acid Battery Sulfation

The most frequent cause of battery damage and short lifespan is sulfation, a process in which crystals of sulfur are formed at the lead plates, preventing the required chemical reaction from happening.This is most often due to a low charge or electrolyte level of the battery; therefore, you should always keep your eye on these two important features.

Monitoring a Flooded Lead-acid Battery

The two above-mentioned parameters should always be monitored, especially in a flooded lead-acid battery.

The following tools are required for monitoring and refilling a flooded lead-acid battery: distilled water, proper safety gloves and goggles, voltmeter (to figure out the voltage of the battery) and temperature compensating hydrometer (to determine the specific gravity of the battery).

Frequent maintenance and refilling are required to maintain the electrolyte level and the charge level of the flooded lead-acid batteries. In AGM lead-acid batteries, these two factors cannot be modified, meaning that AGM lead-acid batteries are less prone to damage.

As mentioned above, two factors have a significant impact on a battery's health: fluid level and state of charge.

- **Fluid level**

    By opening the flooded lead-acid battery lid, you can check the mineralized (distilled) water level. While most batteries have a fill level, you can make sure to have enough water if no metal lead surface is visible. Make sure you never overfill the battery; the maximum amount should usually be half an inch below the cap. Always consider safety measures mentioned in the previous chapters. The acid, like many other electrical components can be seriously harmful.

- **State of charge and level of discharge**

    By checking the voltage and specific gravity, we can quite simply figure out the depth of discharge and the state of charge of a lead-acid battery.

The following table contains the figures for depth of discharge in relation to voltage and specific gravity for a 12-volt battery. You can easily double the figures and use them for a 24-volt battery as well.

State of Charge	Specific Gravity	Voltage (12 V)
100%	1.26	12.7
75%	1.22	12.4
50%	1.19	12.2
25%	1.15	12.0
0%	1.12	11.9

Charging Considerations

There are three main phases of charging that are important to know before setting up your battery:

Float or trickle charging is meant to charge the battery at the same rate as discharging it and will keep the battery fully charged.

Bulk charging is needed when charging a fully discharged battery. The voltage goes high to reach the maximum recommended voltage.

Absorption phase which follows the bulk phase where voltage is kept at its maximum, while amperage is decreasing until the battery is fully charged.

Best results are achieved with adjustable charge controllers programmed for the mentioned phases and according to the battery's manufacturer recommendations.

In small off-grid systems, the charging period is considerably short for bulk and absorption charging; therefore, you can easily set the same voltage for both phases.

Temperature Effect on Lead-Acid Batteries

Lead-acid batteries provide their optimal performance if stored under optimal temperature (25°C, or 77°F). In general, lead-acid batteries get charged faster, and so, unfortunately discharge faster in higher temperatures, while in lower-than-optimal temperatures, the battery has a decreased capacity, although an increased lifespan. Generally speaking, an 8°C (15°F) rise in temperature will lead to halving the lifespan of a lead-acid battery.

To charge a lead-acid battery in higher and lower temperatures than are optimal, special considerations must be taken into account, the most important of which is the voltage adjustment.

These batteries need to be charged with a higher voltage in colder weather, and a lower voltage in hotter weather.

Added to that, when lead-acid batteries get colder, their capacity decreases, while load and charge rate are affected as well. These factors can seriously affect your storage capacity.

There would also be a risk of higher currents and, consequently, a fire hazard in higher temperatures;

therefore, certain consideration must be taken into account to size your battery bank and your system as well.

Peukert's Law

In colder temperatures, the capacity of lead-acid batteries is reduced. This is explained by Peukert's law. Considering the battery's internal resistance and recovery rate, Peukert's law has formulated an exponent that indicates a battery's actual discharge time. The smaller the number is, the longer the actual discharge time. Lead-acid batteries have a Peukert's number range between 1.3 and 1.5. Let's see how it works:

A 100-amp-hour battery with a 20 amp discharging rate and a Peukert's exponent of 1.3 gets fully discharged in (100-amp hour/20 amp =) 5 hours; however, to know the actual discharge time, you need to also divide it by Peukert's exponent- in this case 1.3 (5 / 1.3 = 3.8 hours, the actual discharge time). Additionally, temperature can affect these figures. The picture below shows how higher temperatures increase the capacity, as well as decrease the C rate (discharge or load factor).

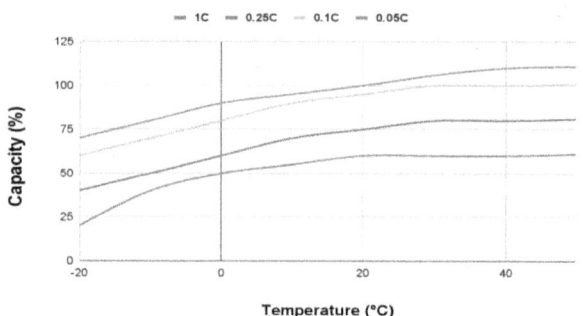

Temperature can be harmful to the battery and the batteries' lifespan, as well as posing a fire hazard. As the temperature goes high, lower voltage is required for the given charge current. As voltage is reduced, the current goes higher to compensate and to keep the equation constant:

V = I * R

The increased current, together with the heat generated from the charging process, may lead to a situation called run-away condition, which can create fires if not controlled.

Due to different manufacturing procedures, various lead-acid batteries seek different modifications, especially in regard to charging voltage. You can easily find the amount of temperature compensation voltage in the specification sheet of lead-acid batteries.

You need to program your charge controller according to the temperature compensation voltage and the local temperature where the battery is located. As explained in chapter six, it is expressed as millivolts per ^{0}C per cell. Do not forget that every cell consists of two volts; therefore a 12-volt battery is composed of (12 volts/2 volts=) 6cells.

Alternatively, if constant charging current is desired, a reduction by approximately 3 millivolts per Celsius degree below 25°C is recommended.

Two common lead-acid battery types (VRLA and FLA) require different charging and maintenance considerations. Below, you can see the main differences regarding the maintenance of these two battery types.

VRLA vs. FLA Battery Charging and Maintenance

Flooded, unsealed lead-acid batteries need the most maintenance of all lead-acid battery types. They need to be frequently filled. In regards to ventilation, they need a properly ventilated space, and it's a good idea to connect the battery box to an outdoor space. In larger systems, ventilating fans are needed to be installed. As explained before, this is due to the hydrogen-releasing nature of these kinds of batteries, especially when overcharged.

VRLA batteries, however, can never be refilled because they are sealed by the manufacturer. Since they only release gas if overcharged, ventilation consideration is still needed, although not as strictly as for FLA batteries.

The level of discharge of the FLA batteries should be determined by measuring their specific gravity via a hydrometer; however, in the VRLA type (and AGM and GEL), this is impossible due to their sealed nature. FLA batteries accept the widest range of voltage tolerance among lead-acid batteries. Accordingly, an appropriate charge controller must be connected to them with a proper charging program.

VRLA batteries are much more voltage-sensitive when compared to FLA batteries. To have a fairly long lifespan, they require proper charging. At higher voltages than recommended, the battery heats up, and gas is released, which leads to permanent harm to the battery.

Hydrogen release in this situation poses a risk of irreversible damage to the battery, while in FLA, it is much less risky. Selecting a proper charge controller is the key to having a well-maintained VRLA battery.

As their name implies, flooded (both sealed and unsealed) lead-acid batteries cannot be placed on their side due to the risk of leakage. VRLA batteries can be oriented on their sides, thus providing more flexibility for battery placement and installation.

Lithium-Ion Battery

With the growing popularity of electric vehicles, lithium-ion technology has been found to be incredibly helpful and reliable in regards to storing electrical energy. These batteries are also now being commonly used in almost any cordless electrical device, such as laptops, cell phones, etc.

Tesla Powerwall and LG Chem are the pioneers of the lithium-ion battery technology utilized in residential solar storage systems.

While lead-acid batteries are more suitable for small, infrequently-used off-grid systems, lithium-ion batteries best suit larger systems.

Longer lifespan and depth of discharge, higher efficiency, and capacity makes lithium-ion batteries an ideal option for residential purposes. Lithium-ion batteries are considerably more expensive than lead-acid. The higher cost, along with the heat-releasing chemistry of the battery, are their only downsides. The heat-releasing feature of these batteries, called thermal runaway, can highly increase the chance of a fire initiation, especially if installed improperly.

Lithium-Ion Battery Mechanics

Any lithium-ion battery consists of the following components:

1. Cathode
2. Anode
3. Electrolyte
4. Separator
5. Two negative and positive current collectors

While lithium ions are stored in the anode and cathode, they are carried from the anode to the cathode and vice versa via the electrolyte and through the separator. This movement will result in releasing free electrons in the anode, hence a charge at the positive current collector. The electrical current will flow from the current collector through the electrical device being powered by the battery to the negative collector. The separator's task is to block the flow of electrons within the battery.

Comparing Lithium-Ion and Lead-Acid Battery

- *Cost:* while you can purchase a 100-Amp hour, 12-Volt lithium-ion battery for around one thousand US dollars, a same capacity VMAX lead-acid battery is worth around one-fourth of the lithium-ion battery. Considering the 50% discharge limit of lead-acid batteries when compared to 100% in the lithium-ion type, you will need two lead-acid batteries to store the same as one lithium-ion, which would still cost you half the price of a Lithium-ion battery.

- *Efficiency*: lithium-ion batteries are proven to be more efficient than lead-acid ones. While an overall 95% efficiency is considered for lithium-ion batteries, the lead-acid type has a range of efficiency between 80 to 85%. Higher efficiency means that a greater amount of stored energy can be provided as usable.

- *Capacity and depth of discharge:* because of their modern technology as well as their chemistry, lithium-ion batteries store more energy than their older competitor with the same amount of space. Overall, this means less space is needed. This will be a decent feature, especially in larger off-grid systems where batteries occupy a substantial amount of space.

  As explained above, depth of discharge is defined as the percentage of battery storage being safely utilized before harming the battery. While lead-acid batteries can be seriously damaged if drained by more than 50% of their total capacity, lithium-ion batteries provide 85% and even more of their capacity ready to be used.

Effective capacity is the term used to describe the depth of discharge joined with the capacity of the battery.

- *Life span:* lifespan of batteries is defined as the number of cycles they can be charged and discharged before becoming seriously damaged. Lithium-ion batteries last several times longer, leading to a more effective lifespan than the lead-acid ones.

Deciding on Battery Type

Bearing in mind all that's been said, there is no doubt that lithium-ion batteries are worth every cent because of all the features we've explained; however, for small systems and for those beginners who just want to wet their feet in the solar system, lead-acid batteries are the first choice. Their fairly low upfront cost makes lead-acid batteries the perfect choice for low-budgeters.

Less frequently used RVs, boats, and cabins can also benefit from economic lead-acid batteries.

Residential battery-backed systems can benefit from all the above-mentioned advantages of lithium-ion battery technology, particularly if a more cost-effective and space-efficient PV system is intended.

Chapter Summary

Batteries are increasingly becoming an inseparable part of solar power systems. As introduction of new battery technologies has made them more cost-effective than ever, having the proper knowledge of different battery types and functions will help you make the most out of your PV system.

In the next chapter, we will discuss inverters and give advice on how to handle them.

DIY SOLAR POWER FOR BEGINNERS

About DIY SOURCE BOOKS

Here at DIY SOURCE, we create and publish paperbacks and eBooks on a variety of topics related to DIY projects. In cooperation with contractors, engineers, electricians, and specialists in various sectors, our authors provide easy-to-use companions for your DIY projects.

Having years of experience in writing, creating, and publishing books, we realized that there are two main reasons why readers do not enjoy reading instructional books:

1. Many instructional guidebooks assume either their readers have a fairly deep knowledge on the subject, and so, confuse readers with a complicated, theoretical, or highly technical manuscript, which will result in ignoring most essential, practical details.

2. Others consider readers as dummies who just want to spend an afternoon reading about a random topic. They spend several pages to describe how to use a measuring tape for instance, instead of telling them that measuring tape is simply needed to be used.

We are highly specialized in creating instruction-based books without unnecessary, repetitive topics that are used as the filler for the books.

Each of our books is designed to teach you what you require to know in order to facilitate the job, fix an issue and complete your project as quickly and accurately as possible. We have avoided unnecessary material, only providing practical, useful information that allows you to master at your chosen subject as soon as possible.

Endlessly trying to improve our books in order to allow anyone to learn new lessons and to practice them easily is not only our job, but has also turned out to be our PASSION.

Although our instructional books are designed in a beginner-friendly manner, professionals can still learn a lot from our books. We teach lessons in a step-by-step fashion and use numerous real life examples in order to prepare our readers for more complicated topics.

We hope this manual help you save time and money while getting prepared for your DIY Solar Power System design, installation, and maintenance.

DIY SOURCE BOOKS

CHAPTER EIGHT: Solar Inverters

A solar inverter is considered the brain of PV systems. As the generated power by the PV module is in the form of DC, an inverter is an inevitable part of any PV system. This chapter is dedicated to inverters to explain the essential features of all types of inverters. If you want to learn more about a PV system, keep reading.

Inverter's Function

The main function of an inverter is to convert the DC load to AC load. There are several different types of inverters available; a grid-tied inverter is connected to the solar array and delivers AC load to the house, while a hybrid inverter is connected to the battery on one side and the grid as well as the AC appliances on the other side to deliver AC load to the house and the grid. Inverters may also provide DC power to the battery.

An inverter/charger can charge your battery using either grid or generator. It can really be useful in off-grid systems where less peak sun hours are available.

Inverter Technologies:

Modified Sine Wave (MSW)

Traditionally, the inverters created a modified sine wave, meaning that the generated electricity had modified steps to mimic the sine wave. Total harmonic distortion (THD) is the metric used to measure the smoothness of the curve, which is measured by an oscilloscope. The lower the THD is, the purer the wave will be.

MSW inverters are way less expensive, but can harm your appliances, especially if you have inductive loads such as refrigerators and pumps. They have a THD of around 30%.

Pure Sine Wave (PSW) Inverters

PSW inverters are more expensive and are the most common ones used in most PV setups. They generate a THD value of less than 3%, which is even smoother than the grid's; with a THD that is less than 5%.

Types of Inverters by Location

- **String/centralized inverter:** this is the traditional, most reliable, and cost-effective inverter, especially for simple roof types with minimum shading. This centralized inverter works at the level of the string(s); therefore, shading of even one panel in the string will result in significant inefficiency in power production. This chapter focuses on this type of inverter.

- **Power optimizers:** each panel is connected to a power optimizer to regulate the DC power output on each PV module and then send it to the central inverter to convert it to AC power. This addition will slightly raise the costs. In fact, this is not a real inverter because it does not convert the DC power to AC.

- **Micro-inverters:** With each panel connected to one micro-inverter, the power is converted to AC on the roof and at the panel level. It ensures there is smooth operation even when there is shading of panels. Although there is no need for a central inverter, using micro-inverters will significantly raise the upfront costs.

Common Features of Inverters

- *DC and AC Terminals*

All inverters should have a DC terminal to connect to the battery, as well as one or a few AC terminals to be connected to the loads.

As the DC terminals should be connected to a large size wire (due to the battery's high current), it is preferred to be made of a fairly high-quality material. Loose connections can lead to significant issues. Copper wires and wire logs are used to connect the battery to the DC terminal.

The AC terminals may be in the form of a regular receptacle, a GFCI receptacle, or a terminal block in smaller inverters, while it is only offered as a terminal block in larger inverters. AC terminals in most inverters have an overcurrent protection device to protect against high amps.

AC terminal blocks are useful for connecting panels and subpanels to the inverter, while power outlets will let your appliances be directly plugged into the inverter.

Regardless of the inverter size, the power should not exceed 1500 watts per each power outlet.

- *High Voltage Disconnect Switch*

Most inverters have a switch to protect against higher voltages than recommended. A 24-volt inverter should only be connected to a 24-volt battery bank; however, if the battery voltage goes high- for instance, 33 volts- the switch will disconnect the inverter. Keep in mind that the switch does not guard this inverter if it is connected inadvertently to a nominal 48-volt battery.

- *Low Voltage Disconnect Switch*

This feature comes in handy, especially if a lead-acid battery is used. This will protect the battery from over-discharging, and consequently, irreversible damage.

A programmable low-voltage disconnect switch is another feature that only some inverters offer. In the case of using a BMS and lithium-ion battery, this feature allows you to turn on the inverter again once the battery voltage rises to the preset voltage.

Voltage and Power Ratings (Wattage)

Inverters are rated according to their voltage. They can be connected to 12, 24, 36, or 48 volt systems. Inverters must match the nominal voltage of the corresponding PV system; otherwise, they will get damaged in the long run.

Inverters are generally available in different power ratings. You may find inverters as small as 50 watts and those as big as 50000 watts; however, most residential inverters utilized in common PV systems are rated between 3000 and 12,000 watts.

Depending on the system's battery size and amount of power generated, inverters are selected to handle your specific power requirements. These components are rated for two different power requirements:

1. Continuous (typical) Power

Continuous power refers to the amount of electricity needed to keep all your appliances working, such as the TV, laptop, fridge, etc.

In fact, continuous power is defined as the amount of power an inverter constantly provides, and is usually much lower than the peak power. This is the amount of power, for instance, a microwave or refrigerator consumes constantly after the motor has started up.

2. Peak (surge) power

Other than continuous rating, inverters are rated for a peak power. Peak power is defined as the maximum power that should be supplied for only a short period of time—a few seconds to minutes. Some appliances, especially those with electrical motors- such as refrigerators, pumps, and compressors- need a large amount of power to start up. This amount of power is much higher than what they consume continuously.

The rate of this surge power can range from 30% to 300% of the inverter's previously-mentioned continuous power. This rating and the time (in seconds) available are mentioned in the inverter's datasheet. Typically, 5 to 15 seconds of peak power rating suffices the power needed to start most appliances. Many household appliances and water pumps might actually need this surge for less than a second.

The example below explains the two different power ratings of an inverter:

If you are running two 100-amp hour, 12-volt batteries, you need an inverter to handle 2400 watts:

2 * 12 volts * 100 amp = 2400 watts

So, a 3000-watt **growatt inverter** can handle your system. This inverter has the ability to handle 3000 continuous watts as well as 9000 surge watts.

Purchasing a larger inverter size than your existing consumption is a good idea. You may need to add more appliances to your house in the future.

Inverter's Cooling System

Inverters generate heat, so they need to be cooled. This is done by using fans on one side and air intake on the opposite side.

A metal heat sink covers the body of the inverter to radiate the generated heat. The radiating heat sink, the fans, and the air vents should never be covered.

Some inverters have recommendations that they can be placed horizontally; however, vertical placement where a fan is located on top and cold air intake downward is preferred to ease the air circulation.

Inverter monitoring

Since inverters may be located far from your reach, a remote switch that can be connected to the inverter is a useful feature. The switches will help you avoid battery discharging and reduce the inverter's idle power consumption when all loads are switched off. You can, alternatively, use a more expensive **Victoran inverter, for instance,** with very low standby power consumption and leave it on for the whole day.

Idle or Standby Power Consumption

This is referred to as the power used by inverters as a load. It may vary between 10 to 60 watts, depending on the quality of the inverters. You can find the standby consumption of any inverter in its datasheet; however, most inverters have a power savings mode that may switch to standby when not a substantial load is connected to them, thus significantly lowering the amount of idle consumption.

If you consume a great amount of power continuously all day, a more expensive inverter with approximately 10 watts of idle consumption better fits your system; otherwise, less expensive ones with higher idle consumption can be an appropriate choice. Do not forget to consider the inverter's idle consumption while calculating your daily electric needs and consumption.

Inverter Efficiency

The conversion efficiency is defined as the amount of AC power output when compared to the inverter's DC power input.

Typically, modern inverters have an efficiency rate of between 80 to 90%. A 90% inverter is considered a premium option to get the most out of your solar array's power generation. Also, if your total load consumption is substantially high, it is a good idea to invest in a high-efficiency (93% and higher) inverter to save a considerable amount of power.

Inverter/Charger

Some inverters have the ability to function as chargers, as well. They can charge batteries via grid or a generator. Some can even use generators to feed your household appliances while a considerable amount of load is being used.

Let's dive deep into the most important topics about inverters:

Types of Inverters

There are three main types of inverters available, according to the type of PV system installed:

1. Off-grid inverters

2. Grid-tied inverters

3. Hybrid inverters

Let's start with off-grid inverters:

Off-Grid Inverters

Off-grid inverters are available in 12, 24, 36, and 48 volts, and should match your solar array and your battery.

Although some inverters possess a protection voltage system, there is no way to use a lower voltage inverter with a higher voltage battery, if you don't want your inverte to be permanently damaged. Other than voltage, inverters are rated in watts. The bigger your battery bank is, the bigger the inverter is needed to provide power to your AC.

Grid-Tied Inverters

Grid-tied PV systems need inverters. The inverter is regarded as the brain of the system by managing the flow of power. The main role of the inverter is to convert the DC power generated by the solar panels into usable AC power for AC appliances in your home.

As its name implies, DC power flows in one direction; AC power, however, alternates due to changing the direction of the power. More efficient inverters generate more AC power out of the PV system's generated DC power.

In the US, grid-tied inverters are required and mandated by the National Electrical Code (NEC) to be installed in grid-tied PV systems, as explained in previous chapters.

With net metering, the local utility company pays or charges for the net electricity transferred into and from the grid by a grid-tied inverter. This is recorded by the specific meter on the customer's premises.

For instance, if you consume five kilowatt-hours per month and your PV system generates and transfers four kilowatt-hours to the grid per month, you will be charged by your local utility company just for one kilowatt-hour balance of the electricity transferred to and from the grid. In the US, net metering policies may be different state by state.

The grid-tied inverter should be able to match the phase of the local grid and keep its power output voltage higher than the grid's voltage. This higher voltage would help to transfer the generated power to the grid.

Most modern grid-tied inverters provide a fixed unity power factor. This means that the inverter's output voltage and current's sine waves are perfectly in phase, and the phase angle may differ only one degree from the grid's AC power.

The most significant drawback of these inverters is that they are required by the NEC to switch the whole AC and DC circuits off in the event of a power outage. This is for the safety of electricians who are fixing an issue following the blackout.

Hybrid Inverters

Hybrid inverters are considered the headquarters of battery-backed PV systems; they are, in fact, a combination of a regular grid-tied solar inverter and a battery inverter/charger in one unit.

These inverters use meters to measure electricity consumption and smart software that is programmed to determine an efficient way of using and converting solar energy.

As the description above implies, hybrid inverters perform multiple functions; therefore they have some limitations when compared to off-grid inverters. The main limitations are as follows:

1. Limited peak (surge) power output when a blackout happens; therefore, appliances that need peak power to start are not usually connected to them.

2. Most hybrid inverters have limited backup power; so, only small (essential) loads such as lighting or some other low-consumption devices can be backed up when the grid is down.

Unlike grid-tied inverters, hybrid ones let you still use the power for essential appliances in the event of a blackout.

AC-Coupled and DC-Coupled Configurations

AC- and DC-coupled configurations are battery-related topics; however, the inverter's integration in the two configurations is highly different, and you need to understand the main differences to choose the proper one for your house.

It is a specific feature of battery-backed systems, and you need to select the suitable configuration and the corresponding inverter type accordingly.

As you know, solar array produces DC electricity, while most household appliances utilize AC power; however, the battery bank stores the power in DC form.

The main difference between the two systems is the path that the generated DC power should take to be stored in a battery, converted by an inverter, consumed at home, and transferred to the grid. Let's dive into more details of each system.

We will discuss both types of configurations and their specific use here:

AC-Coupled System

In this system, the solar array's generated electricity flows to a regular solar grid-tied inverter to be converted to AC electricity.

Depending on the time of the day and your energy consumption, the AC power can be used for your home appliances, sent to the grid, or go to an inverter/charger to be converted back to DC to charge the battery bank.

The same inverter/charger mentioned above should convert the battery's DC power to AC for use during night or blackouts.

In this type of configuration, the stored power in the battery should be converted three separate times to be ready to use. This configuration will eventually result in lower efficiency rate when compared to DC-coupled system.

AC-Coupled System

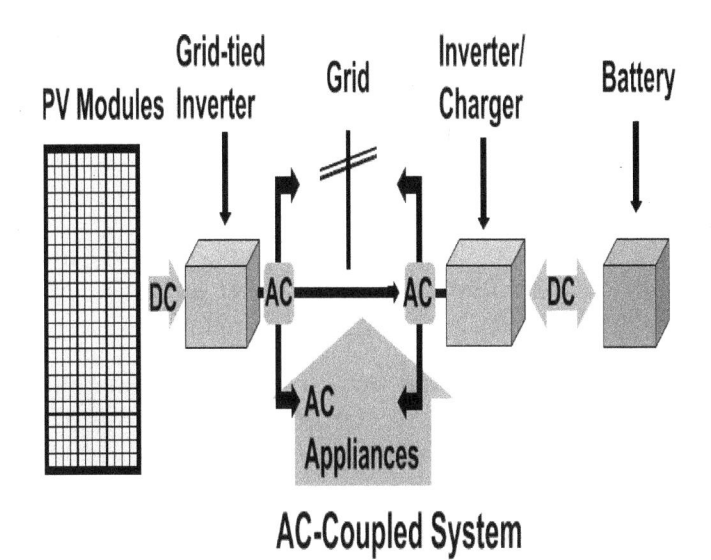

As you can observe in the diagram above, no charge controller is installed, while two inverters are needed in the AC-coupled configuration. Inverters carry out the following tasks in this configuration:

- *Grid-tied inverter*

This inverter converts the DC power from PV modules to AC power in order to be usable by the AC appliances and the utility grid. This inverter then decides to direct the power toward the grid, your house, or the second inverter, based on your power consumption and its preset program. Note that the grid-tied inverter cannot feed your battery bank. When PV system's generated power surpasses the daily consumption, this inverter will redirect the electricity toward the second inverter or the grid depending on how it has been programmed.

- *Inverter/charger*

This smart device is simultaneously connected to the grid-tied inverter, the grid, your house, and the battery system.

Once the batteries are fully charged, this inverter sends a signal to the grid-tied inverter, meaning that all the converted AC power can either go to the grid or to the house, based on your consumption.

This inverter can even use the grid to charge the battery if your array's generated power does not fulfill your battery capacity.

This type of configuration is the most desired one for those homeowners with an existing grid-tied PV system who decide to add a backup to their system.

Pros of AC-Coupled System

1. Ease of installation, less labor and time, and a lower upfront cost, especially for already-existing grid-tied systems.

2. Batteries can be charged from both the solar array and the grid, especially for resiliency or electricity rate arbitrage benefits.

Drawback of AC-Coupled System

Stored solar electricity should be inverted three times before getting ready to use, which will result in the reduction of the system's efficiency.

DC-Coupled System

In this type of configuration, the generated DC power flows to a charge controller to charge the battery bank similar to a totally off-grid system; therefore, the power is not converted to AC and then back again to DC to be stored in the batteries. A hybrid inverter performs all the tasks that two inverters do in an AC-coupled system. This inverter converts the battery's DC power to usable AC power for your home and the grid. It can also use the grid to charge your battery bank.

There are two main paths that the array's generated power will pass through, and in both ways, the electricity will be converted only once from DC to AC. These paths include:

1. The power flowing from the battery bank to your home

2. The power directed from battery to toward the grid

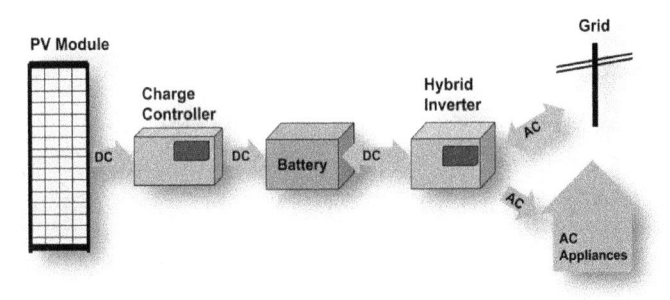

DC Coupled System

Historically, AC-coupled configurations were more common for residential and commercial solar installations since most home owners started with a conventional grid-tied system and then added a battery bank to it; however, as more DC options are available now, DC-coupling is getting more and more popular.

Advantage of DC-Coupling System

They are more efficient due to one-time power conversion.

Drawback of DC-Coupling System

They are more complicated to install, which can add to your upfront costs, as well as installation time.

How to Size an Inverter

One of the most important steps to select the appropriate inverter for your PV system is to figure out the correct inverter size needed. Since certain PV systems and inverter types have specific sizing requirements, we will discuss the sizing for each type of inverter separately.

Size an Off-Grid Inverter

As previously explained, an off-grid inverter is sized according to your daily continuous and surge power consumption (wattage). Unlike sizing the battery bank, where you need to figure out your total energy consumption (watt-hours), sizing an off-grid inverter is merely based on your consumed power (wattage) per day. In other words, your off-grid inverter should be able to provide power so that all your AC appliances can run smoothly regardless of their duration of usage per day. This example will help you learn how to size your own inverter. The chart below represents a list of loads, including the continuous and surge power consumption of several appliances used in an off-grid house. A simple alternative way of sizing is using an **online calculator.**

AC appliance	Quantity	Continuous Power	Surge Power
Light bulbs	10	100	**0**
Fridge	1	250	2200
Microwave	1	1000	0
Well pump	1	425	2350
LED TV	1	175	0
Coffee maker	1	950	0
Total(watts)		2900	4550

As the table implies, we need an inverter that can handle at least 2900 continuous watts, as well as 4550 surge watts for some appliances to kick-start. Additionally, the voltage should match our nominal battery voltage.

Other than the maximum power, the inverter must match your AC appliances' voltage. Inverters are available in 12, 24, and 48 DC volts. The DC input voltage of an off-grid inverter must comply with the battery's nominal voltage; otherwise, it may not work or may be damaged. The AC output of inverters may be 120 or 240 volts, depending on the inverter type and the output wiring.

Size a Grid-Tied Inverter

To size your required grid-tied inverter, you need to look for two important features:

1. Power (wattage)
2. Array's voltage

As we discussed before, a grid-tied inverter's main function is to receive and convert the generated DC power.

The inverter must be big enough to handle the maximum amount of power generated by your solar array.

Assuming you have installed two strings of solar panels- each composed of four solar panels rated at 175 watts, 12 volts- let's figure out the required inverter's maximum power by doing the following simple calculations:

Maximum power=2(parallel strings)*4(panels in every string)*175 = 1400 watts, total solar array's generated power

So, a 2000-watt inverter can easily handle the generated power.

Since you may need to add more panels to cover a greater portion of your electricity bill in the future, it is wise to oversize your inverter.

It avoids re-doing and the need to replace your inverter with a larger one in the future.

Chapter Summary

Inverters are regarded as the decision-making center for your PV system. By selecting the correct inverter, you will secure efficient power production and consumption, as well as storing and transfer of power to the grid.

The next chapter will focus on the wiring of PV systems. By the end of the next two chapters, you will be able to connect all the components properly and get ready to turn on your PV setup.

DIY SOLAR POWER FOR BEGINNERS

CHAPTER NINE: Conductors and Connectors

If you're installing a PV system in your home, you need wires and cables to connect solar panels to the rest of your PV system. A wire consists of a single electrical conductor, while a cable consists of a group of wires. Cables are used to carry the electrical current from one solar component to another.

Choosing the right type of wires is critical to the proper functioning of your photovoltaic system. Without proper wiring, power from panels may be unnecessarily lost, hence less power to fully charge the battery. Since most of the solar installations are located outdoors, the wiring has to meet tough environmental requirements, such as moisture, heat, and UV resistance.

Composition of Wires and Cables

Generally, wires are either single or stranded *type*. Wires with a single metal wire core have a single-stranded conductor. Wires that have multiple wire cores form a multi-stranded conductor.

Single or Solid Core Wires

A solid wire has single-core wiring that can be insulated or buried. This type of wire is used to connect two wiring centers, or to connect a wiring center and a conjunction box. Solid core wires have a compact diameter, making them more cost-effective. Single metal conductor wires use protective sheath insulation and are great for wiring indoor solar systems.

Stranded Core Wires

Stranded wires have several conductor wires twisted together to form a multi-stranded wire. The wires are enclosed into a protective sheath or insulator. Multi-stranded wires are available in large diameters and have the ability to handle frequent movements.

Due to the high attenuation of these stranded cables, you can connect solar components within short distances. If you live in areas that regularly experience extremely high winds, you should go for multi-stranded conductors because they are more flexible and durable. Multi-stranded conductors perform better and are ideal for connection in cars, boats, and RV vans.

Types of Solar Wires

Two of the most commonly-used wires for solar installation include:

USE-2 Wires

A USE-2 wire is appropriate for connecting the array's terminals in grounded systems. The wires are rated to handle up to $90^0 C$ of wet and dry weather conditions.

This type of wire is ideal in places with minimal movements of the wire, and can be made out of solid or stranded conductors.

PV wires

This type of wire interconnects PV modules and can work well in areas that experience 90^0C heat and moisture. They can also handle dry conditions of up to 105^0C. This type of wire has thick insulation and is made of a single-stranded wire core. The wire is less flexible and resilient.

The insulation jacket of PV wires offers superior sunlight resistance, a proven level of flame resistance, and low-temperature resistance. PV wires are used in ungrounded PV arrays.

Conductor's Main Features

Before selecting the right wire size and type, you need to know the most significant wire features.

Wire Ratings

All wires are rated by amps (amperage). This is the maximum amount of current that can travel through the wire. The higher the amps the system is rated at, the thicker the wires needed. Solar cable manufacturers publish rating charts that indicate the number of amps (current) a particular cable can safely handle. Wires that have higher ratings can handle higher amps and have a lower voltage drop. They also have less risk of overheating.

For example, if your system generates 10 amps of current, then you need at least a 10-amp rated wire to handle the current. If you use a wire rated at a lower amp than what the panels generate, it will result in a voltage drop and the wires are likely to heat up.

Wire Material

Solar wires are also classified based on the construction material. The wires are made of either copper material or aluminum. Those made of copper material are more conductive and carry more amps than aluminum wires.

Though copper wires are expensive, they're highly recommended because they offer better flexibility and heat resistance. On the other hand, aluminum wires are prone to damage in extreme temperatures and are more vulnerable when bent.

Wire Insulation

Different types of solar wires have different insulation material that protects the wires from heat, moisture, water, UV light, and chemicals. The standard insulation types include:

- ***THHN***: this protective sheath works well on applications installed in dry areas or indoor conditions.

- ***TW/THW/THWN***: this is suitable for conduit wires installed in wet areas and in both outdoor and indoor conditions.

- ***UF & USE*** (underground service entrance): the wires have insulation sheaths for protection on wet surfaces, and for underground wiring.

- ***THWN-2***: this protective sheath is used for indoor wire installation and is less expensive. Since the wire should run through a conduit, the sheath is not UV resistant.

- **RHW-2**: this type of insulation is mostly used on *PV wires* and *USE-2 wires*. The insulator works well in moist areas and is ideal for outdoor applications. The PV wire and USE-2 jackets can protect against extreme UV exposure and are moisture-resistant. PV wires have an extra layer of insulation as well.

Wire Color Codes and Solar Application

Color-coded solar wires make it easy to map out the electrical wiring of your solar system. These wires are color-coded to designate their function and uses. Wire colors are essential for troubleshooting and repairing your solar system.

Both AC and DC applications use different color codes. The color guide for a simple AC and DC application includes:

AC		DC	
Color	**Application**	**Color**	**Application**
Red, black, or other color	Un-grounded hot applications	**Red**	Positive pole
White	Grounding conductors	**White**	Grounding or negative
Bare or Green	Equipment grounding	**Bare or Green**	Equipment grounding

You should always follow the recommended National Electrical Code (NEC) guide when installing your solar system.

If you don't know the specific conductor or insulation of a certain application, you can seek professional assistance.

Below, you can see the major applications of wires in PV systems:

DC Solar Cables

DC solar cables are single core copper wires with sheath insulation to handle high temperatures, high UV radiation, and they are weather-resistant. These cables are prebuilt into the solar panels and they come with connectors.

PV Output Cable

The main DC cables connect the positive and negative terminals of panels from a junction box to the load center of the PV system. They come in different sizes of 2 mm, 4 mm, and 6 mm. Main DC cables can either be single or two-core cables (negative and positive). The single-core wires have double insulation, making them highly reliable.

AC Connection Cables

AC connection wires connect the solar inverter to the electrical grid or other protective equipment. If you have a small solar system, you can use 5-core AC cables to connect to the electrical grid. In this case, you should connect three live wires to carry electricity, another wire for grounding, and a neutral wire.

Sizing Solar Wires

When selecting the wires, you have to choose the correct gauge.

The most common wire scaling system is the AWG—American Wire Gauge. Wire gauge determines the amount of current intensity that flows through the circuit.

Wires with lower gauge numbers have lower resistance and, as a result, a high amount of current will run safely through the wires.

When sizing your solar wires, you need to consider the following:

- PV module and inverter spec sheet.

- Array's configuration (how many panels to configure in series or parallel).

- Whether you need a junction box or a combiner box.

- Voltage of the circuit and voltage drop index.

The following table indicates the capacity of various wire gauges and normal residential usage designated by the American Wire Gauge.

You can use this table as a basic tool to roughly estimate the right wire size for your PV system. To be more accurate, you need to exactly follow the National Electrical Code (NEC). Consider this table as a warmup to later dive deep into NEC's tables.

Wire Gauge (AWG)	Amperage	Application
3/0 gauge	200 amps	Service entrance
1/0 gauge	150 amps	Service entrance and feeder wire
3 gauge	100 amps	Service entrance and feeder wire
6 gauge	55 amps	Feeder and large appliance wire
8 gauge	40 amps	Feeder and large appliance wire
10 gauge	30 amps	Driers, appliances, and air conditioning
12 gauge	20 amps	Appliance, laundry, and bathroom circuits
14 gauge	15 amps	General lighting and receptacle lighting

This table provides a rough estimate for your wire size. To be more precise, however, you need to comply with the National Electrical Codes (NEC) when calculating the proper conductor size. The NEC considers three main adjustments for wire sizing. Below, you can read the detailed wire sizing principles and procedures.

Basic Principles of Wire Sizing

When it comes to wiring and connecting the solar equipment, selecting the right conductor size is of great significance. Under sizing the wires for your system may lead to wires overheating and a fire hazard. There are four main considerations to take into account while selecting the right wire size for your grid-tied or off-grid PV system. The first three are mentioned in the NEC; however, the last one is not. These considerations are as follows:

- *Conductor ampacity*
- *Ambient and rooftop temperature adjustment*
- *Number of conductors in a conduit or raceway*
- *Voltage drop*

The wires should be sized based on all or some of these principles for each circuit of the PV system. We will first discuss these principles in four steps, and then we will size the wires of a PV system using all or some of these steps. Due to using various NEC tables, this topic may be confusing, so you may need to reread it until you digest the concept. Let's get started:

Step 1: Wires Ampacity (Amp Capacity) in 30°C

First of all, we need to consider the conductor's ampacity. Ampacity is defined as the maximum current that a conductor can pass continuously under certain conditions of use, while not to a degree that exceeds its temperature rating. You should keep in mind that, wires are temperature-rated at 60°C, 75°C, and 90°C. The NEC table 310.15 (B) (16), or simply *Ampacity* table, has considered all the requirements to determine the maximum current-carrying capacity (ampacity) of the wire. In previous editions of the NEC, the table was numbered 310.16. The following table is a simplified version of table 310.15 (B) (16) or simply the *ampacity* table:

American Wire Gauge (AWG)	75°C-Rated Conductor (Amps) RHW, THHN, THW, THWN, XHHW, USE	90°C-Rated Conductor (Amps) RHW-2, THHN, THHW, THW-2, THWN-2, USE-2, XHHW
16		18
14	20	25
12	25	30
10	35	40
8	50	55
6	65	75
4	85	95
2	115	115
2/0	175	195
4/0	230	260

The nominal ampacity of a conductor at 30°C depends on a number of factors, such as:

1. *Material*, which can be either copper or aluminum. For PV purposes, only pure copper wires are recommended.
2. *Size.*
3. *Insulation type and material.*

4. *Application*, which could be a direct burial, inside a conduit, or in the air.

According to the NEC, while USE-2 and PV wires cannot be used in a conduit, most of the other above-mentioned types should be. You will need to refer to the NEC, ampacity table to match the maximum current (Isc) of each segment of your system's wiring with the required wire gauge or simply use our previously-mentioned table. For instance, you can see that a 75°C rated THHW, 10-gauge wire can carry 35 amps in 30°C. If you do not know your wire's type, just follow your wire's temperature rate (60°C, 75°C, or 90°C) to find the ampacity of your wire in 30°C.

To make the first step short; once you figure out your system's maximum current generated, you can find the smallest wire gauge that can carry the current at 30°C according to the NEC book, table 310.15(B)(16).

Step 2: Ambient Temperature Correction and Rooftop Temperature Adjustment

- *Ambient temperature* can affect the conductor's maximum current carrying capacity (ampacity), meaning that the nominal ampacity of each wire gauge will be reduced in temperatures higher than the standard 30°C and increased when used in colder weather conditions.

According to NEC table 310.15(B)(2)(a), the temperature correction factor for the temperature range of 26°C–30°C is one; for temperature ranges below 30°C, it is bigger than one; and for temperatures above 30°C, it is less than one. You need the correction factor for your local highest temperature. Once you pick the nominal ampacity found on the ampacity table 310.15(B) (16), just multiply it by the correction factor to calculate adjusted ampacity. If you are wiring where on hot days reach 40°C for instance, you need to multiply the ampacity by the temperature correction factor from table 310.15(B)(2)(a). So, temperature-adjusted ampacity = ampacity in 30°C * correction factor of the ambient temperature.

Ambient Temperature (Celsius)	75°C Rated Conductor	90°C Rated Conductor
21-25	1.05	1.04
26-30	1.00	1.00
31-35	0.94	0.96
36-40	0.88	0.91
41-45	0.82	0.87
46-50	0.75	0.82
51-55	0.67	0.76
56-60	0.58	0.71
61-65	0.47	0.65
66-70	0.34	0.58
71-75		0.50
76-80		0.41

For instance, if a 14-gauge wire can carry 20 amps in 30°C, you need to multiply the current by its corresponding correction factor in 40°C (in this case, 0.88) to calculate the actual ampacity of your wire in 40°C (20 amps * 0.88 = 17.6 amps).

As you can see ambient temperature has a significant impact on the wire's ampacity.

- There is another temperature consideration, which is ***the height of the conduit*** on the roof. According to the NEC table 310.15 (B) (3) (C) below, the closer the conduit or raceway to the roof, the hotter it gets. This table mandates that you adjust your wire size according to its height on the roof:

Distance between the raceway and the roof(inches)	Degree Celsius	Degree Fahrenheit
On the roof	33	60
More than ½ inch above the roof	22	40
3 - 12 inch above the roof	17	30
12-36 inch above the roof	14	26

For instance, if you run the wires right on the roof, you will need to add 33°C to your ambient temperature, while only 14°C needs to be added to your ambient temperature if wires are running 12 inches or more above the rooftop. Subsequently, you need to check the table 310.15(B) (2) (a), again to figure out the temperature correction factor. Note that this temperature adjustment only applies to the wires exposed to sunlight and in a conduit; indoor wires or free wires do not require this adjustment.

Step 3: Adjusting Ampacity for Multiple Conductors in a Conduit

PV systems are usually composed of several strings, which consist of a number of individual panels. The total output is called the PV source circuit, which will run into the combiner box located on the rooftop. The PV source conductor(s) should not run in any conduit or raceway.

Wires exiting the combiner box (PV output), however, need to run in an encircled conduit or raceway. In any condition where more than three conductors run in a single conduit or a raceway for a distance that is more than 24 inches, the table 310.15 (B)(3)(a) indicates that the conductor's ampacity should be corrected by a certain percentage as follows:

Number of Conductors in a Conduit	Adjustment Factor (%)
4-6	80
7-9	70
10-20	50

The reason for this adjustment lies in the fact that when more wires are packed in an encircled space, their heat-dissipating ability decreases significantly, leading to hotter wires; therefore, a larger diameter of wire is needed to compensate for the added heat. So once the conductor ampacity is calculated from table ampacity table, it must be then corrected by ambient temperature according to table 310.15(B)(2)(a), and finally, it should be multiplied by the additional conductor correction factor, according to NEC table 310.15(B)(3)(a).

In this table, you can find the corresponding additional ampacity correction factor according to the number of conductors that carry a current. It is 80% for 4–6 wires and 70% for 7–9 wires running in a single conduit. For instance, if a 14-gauge conductor with ampacity of 20 amp in 30°C is to be used for two PV output circuits (past the combiner box) in a conduit (including two positive and two negative conductors), we would need to adjust the temperature rating for four current-carrying conductors. The correction factor from the NEC table 310.15(B) (3) (a) for four wires is 80%. In other words, the above-mentioned wire can only handle 80% of its ampacity when running with multiple wires in a conduit:

20 amps * 80% = 16 amps

When sizing a 3-phase AC system, the neutral conductor is not counted as current-carrying.

Step 4: Voltage Drop

As the distance between the combiner box on the roof and the DC load center may be considerably long, there is a chance of voltage drop. This drop in voltage can result in an increase in the current. We should check if it is significant enough to alter the wire gauge needed. Generally, no extra consideration is needed if the drop is calculated less or equal to 2%.

Using the following online calculator (scan the code), you can calculate your voltage drop. You need to enter the following figures to see if your voltage drop is less than 2%:

1. The wire gauge.

2. Vmp (maximum power voltage) of the array.

3. Imp (maximum power Amperage) of the array.

4. The length of the wire.

windynation	2% Voltage Drop Chart					
AWG =	14	12	10	8	6	4
Capacity(AMPS)	15	20	30	40	55	70
ARRARY AMPS	FEET ONE WAY FOR A PAIR OF WIRES					
1	45	70	115	180	290	456
2	22.5	35	57.5	90	145	228
4	10	17.5	27.5	45	72.5	114
6	7.5	12	17.5	30	47.5	75
8	5.5	8.5	11.5	22.5	35.5	57
10	4.5	7	9.5	18	28.5	45.5
15	3	4.5	7	12	19	30
20	2	3.5	5.5	9	14.5	22.5
25	1.8	2.8	4.5	7	11.5	18
30	1.5	2.4	3.5	6	9.5	15
40			2.8	4.5	7	11.5
50			2.3	3.6	5.5	9
100					2.9	4.6

If the voltage drop is more than that, you need to adjust your wire gauge. An alternative to the online calculator is using the illustrated voltage drop table to adjust the wire length in order to remain within the 2%-voltage drop limit. In the example below, we will use the voltage drop table to see if we need a wire gauge adjustment or not.

If you have a 450-watt, 12 V solar system and a Vmp of 18 V, the Imp (maximum power current) flowing through the panels will be: 450 W / 18 V = 25 amps.

Based on the previously-mentioned ampacity table, the smallest wire gauge you can use is 10 AWG, which is rated at 30 amps, which is more than the 25 amps required.

From the array amps column, the 25 amp row and 10 AWG wire support a wire length of 4.5 feet with a voltage drop of 2%. So there is no need to worry.

Sizing Conductors for PV Circuits

In this part of the chapter, we will use the above-mentioned NEC considerations, as well as the voltage drop calculator to size the wire needed for different circuits of a PV system. These circuits, as you can observe in the following wiring diagram, include:

- PV source circuit
- PV output circuit
- Battery output circuit
- Inverter output circuit

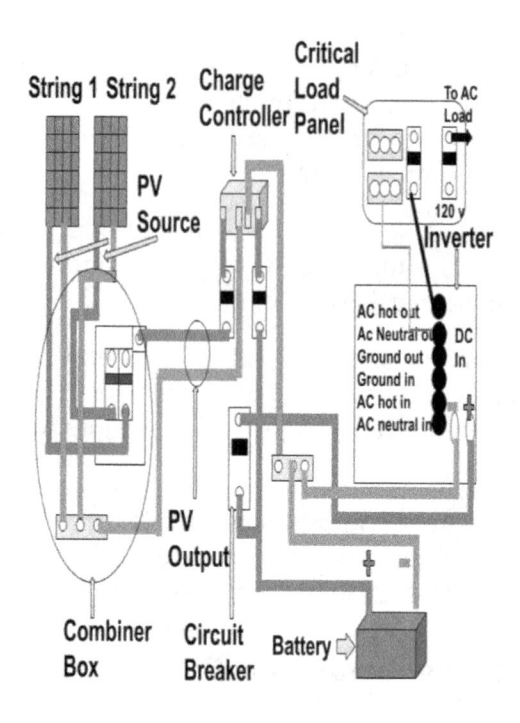

Assume that we want to connect eight 100-watt, 12-volt panels with Isc of 8.5 amps in two parallel strings. We are located in an area where, on the hottest summer days, the ambient temperature reaches 41^0C. Let's now find the smallest conductor gauge needed for each circuit of the system:

PV Source Wire

The PV source, also known as the extension cable, carries the power directly from the panels to the combiner box on the roof. The combiner box contains circuit breakers and fuses, if necessary.

Step 1: Size the Ampacity

To size the correct wire gauge for the PV source, we need to consider the highest amount of current, which is the short circuit current (Isc) generated by the panels.

According to the NEC book, table 690.8(B), all the conductors between the solar array and the inverter must be able to handle up to 156% of the short circuit current (Isc) of each string of the solar panels connected in series. Why?

This is because of multiplying the Isc of panels by 125% twice. As previously explained in chapter Five, the first 125% is considered a safety factor due to more than three hours of continuous use.

As solar panels receive sunlight and generate power all day long, we need to consider this safety factor.

The Second 125% is a safety factor that accounts for the added current of the array due to over irradiance, conditions in which sun exposure may be more than the Standard Test Conditions (1000 W/m^2) , for instance in extremely sunny summer days.

The equation below better formulates what has been explained above:

Ampacity = Isc of each string * 1.25 (over irradiance) * 1.25 (3-hour continuous use)

Assuming that the Isc of each panel is 8.5 amps, so:

Ampacity = 8.5 amps * 1.25 * 1.25 = 13.26 amps

This is the maximum amperage that the wire should handle when the ambient temperature is 30°C.

AWG	75°C rated	90°C rated
16		18
<u>14</u>	<u>20</u>	25
12	25	30
10	35	40
8	50	55
6	65	75
4	85	95
2	115	115
2/0	175	195
4/0	230	260

By referring to the NEC book, ampacity table, we can find the minimum wire gauge (14) that can handle 13.26 amps at 30°C.

As previously explained, this wire will be connected to the wire included in the panels via a connector (MC4, for instance) and will run into the combiner box, with no conduit used. As mentioned before, two of the most common and permitted options are:

- USE-2 for grounded systems.
- PV wires for ungrounded systems.

We will use a USE-2 for this case, which is a less expensive option.

This wire, also called an extension cable, will go inside the combiner box to be connected to a circuit breaker, so the temperature rating of the circuit breaker must be taken into account. For instance, if your wire is rated at 90°C, but the circuit breaker is at 75°C, you need to check the wire gauge for 75°C in the NEC tables, not 75°C.

Additionally, the connector must match the type of the connector you have used to series-connect the panels. Just check the manual of the extension cable to match it with the proper connector (usually MC4).

If panels are supposed to be connected in series only, the watertight junction box and wire included with the panels usually satisfies your wiring needs; however, if strings of panels are to be connected in parallel, which is usually the case, an extension cable with male and female connectors is needed to carry the current from the panels to the combiner box on the roof.

The smallest wire gauge that can handle this much current is a 75°C rated, 14-gauge USE-2 wire that can handle up to 20 amps, which is more than 13.26 amps. Do not forget that this is the smallest diameter. You can use a 12- or 10-gauge wire if a 14-gauge extension cable is not available on the market.

If on the hottest days where you live the ambient temperature would be as high as 30°C, you are good to go; however, if your wires experience higher temperatures, you need to follow the remaining steps as well.

Step 2: Ambient Temperature Correction and the Rooftop Temperature Adjustment

According to table 310.15(B)(2)(A), in the highest ambient temperature range of 41–45°C, we need to consider a correction factor of 0.82 for a 75°C-rated wire. In other words, we should multiply the ampacity calculated using the ampacity table by a correction factor of 0.82 to correct it for the highest local ambient temperature possible.

Ambient Temperature (Celsius)	75°C Rated Conductor	90°C Rated Conductor
21-25	1.05	1.04
26-30	1.00	1.00
31-35	0.94	0.96
36-40	0.88	0.91
41-45	0.82	0.87
46-50	0.75	0.82
51-55	0.67	0.76
56-60	0.58	0.71
61-65	0.47	0.65
66-70	0.34	0.58
71-75		0.50
76-80		0.41

As the PV source (extension cable) runs freely in the air, there is no need to worry about the roof temperature adjustment. So:

20 amps * 0.82 = 16.4 amps in an ambient temperature range of 41–45°C

As you can observe, the ambient temperature can significantly decrease the wire's ampacity; however, in this case, a 14-gauge wire can still handle up to 16.4 amps at 41–45°C, which is bigger than our systems maximum generated current (13.26 amps), according to the NEC, ampacity table. So, there is no need to go for larger diameters.

Step 3: Number of the Conductors in a Conduit

As the PV source (extension cable) runs freely in the air, there is no need to worry about this step.

So we can apply the previously-calculated figure in steps one and two as the maximum amount of current generated, as well as the highest amount of the current that the selected conductors can pass through.

Step 4: Voltage Drop

As the distance between the solar panels and the combiner box on the roof is fairly short, the voltage drop check is not needed here. This voltage drop may happen in the longer PV output conductors, and we will check it there.

PV Output Wires

This wire carries the current out from the combiner box and is usually placed inside a conduit. The same steps for the PV source wire must be taken here.

Additionally, as we have two parallel strings that are combined past the combiner box, we need to multiply the ampacity by the number of strings connected in parallel.

Step 1: Size the Ampacity

Ampacity = Isc of the string * number of parallel strings * 1.56 (over irradiance, continuous use).

So, in our case;

Ampacity=8.5 amps*2*1.56=26.52 amps. As this conductor must run in a conduit, a 75°C-rated, 10-gauge THHN-2 conductor that can carry up to 35 amps at 30°C is the smallest suitable gauge according to ampacity table so far:

AWG	75°C-Rated Conductor (Amps)	90°C-Rated Conductor (Amps)
16		18
14	20	25
12	25	30
10	35	40
8	50	55
6	65	75
4	85	95
2	115	115
2/0	175	195
4/0	230	260

However, we need to check the rest of the steps.

Step 2: Ambient Temperature Correction and Rooftop Temperature Adjustment

For roof temperature adjustment, we need to check table 310.15 (B) (3) (C). If we run the conduit at the level of shingles, we will be required to add 33°C to our ambient temperature. This is due to poor air ventilation around the conduit. If we level it ½ inch higher, only 22°C is needed to be added to our highest ambient temperature (41°C).

Height on the Roof (Inches)	Degree Celsius	Degree Fahrenheit
On the roof	33	60
> ½ inch	22	40
3 - 12 inch above the roof	17	30
12-36 inch above the roof	14	26

41°C + 22°C = 63°C adjusted ambient temperature.

Ambient Temperature (Celsius)	75°C-Rated Conductor	90°C-Rated Conductor
21-25	1.05	1.04
26-30	1.00	1.00
31-35	0.94	0.96
36-40	0.88	0.91
41-45	0.82	0.87
46-50	0.75	0.82
51-55	0.67	0.76
56-60	0.58	0.71
61-65	0.47	0.65
66-70	0.34	0.58
71-75		0.50
76-80		0.41

Going back to NEC table 310.15(B)(2)(A), for ambient temperature correction, we will see that for a

75°C-rated wire, we will need a correction of 0.47 at 63°C, meaning that the ampacity of the selected 10-gauge wire should be multiplied by 0.47 (35 * 0.47 = 16.45 amps). This figure is less than our maximum generated current passing through the PV output wires (26.52). As observed above, the location of the conduit on the roof has a significant impact on the ampacity, and therefore, the wire gauge required. So, we will need a 6-gauge wire, which can handle an adjusted (65 amps*0.47=) 30.59 amps to be bigger than 26.52amps.

AWG	75°C-Rated Conductor (Amps)	90°C-Rated Conductor (Amps)
16		18
14	20	25
12	25	30
10	35	40
8	50	55
_6_	_65_	75
4	85	95
2	115	115
2/0	175	195
4/0	230	260

Step 3: Number of Wires in a Conduit

Since we only have one negative and one positive, and a ground wire, there is no need to worry about table 310.15 (B)(3)(a). If more than three current-carrying wires pass through a single conduit, you will need to check that table. So far, we need to use a

THWN-2, 6-gauge wire in a conduit located a 1/2 inch higher than the roof to the DC load center. It is not finished, though. We need to check the last consideration.

Step 4: Voltage Drop

As the length of the circuit between solar equipment may be long, there is a chance of a voltage drop.

Generally, if the calculated drop is equal to or less than 2%, there is no need for further wire gauge adjustments. When using an online calculator, we need to enter:

1. The selected wire gauge.

2. Solar panel's Vimp (maximum power voltage).

3. Solar panel's Imp (maximum power amperage).

4. The length of the wire between the combiner box and DC load center.

This tool will determine whether the voltage drop is less than 2%. Fortunately, in this example, we are within the 2% voltage drop limit. You can scan the code ain page 226 to use this calculator.

If the voltage drop is more than that, we need to adjust the wire gauge or length however possible.

Battery to the Inverter Wires

Step 1: Size the Ampacity

To find out how much current a cable should be able to carry from the battery to the inverter, we need to know two parameters: the inverter's maximum power (wattage) as well as the lowest battery voltage at which the inverter turns off in order not to drain the battery inadvertently. As Watt's law implies: I = P / V. So, from the inverter's manual or datasheet, you can figure out its continuous power (wattage) and the lowest battery voltage. So, Maximum amperage = inverter's continuous power / lowest battery voltage. Assuming we are using a 2000-watt continuous power inverter with the lowest battery voltage of 20 volts: Highest amperage = 2000 / 20 = 100 amps

AWG	75°C-Rated Conductor (Amps)	90°C-Rated Conductor
16		18
14	20	25
12	25	30
10	35	40
8	50	55
6	65	75
4	85	95
2	115	115
2/0	175	195
4/0	230	260

So far, a 2-gauge wire is the minimum wire size suitable for this situation.

Step 2: Ambient Temperature Correction and Rooftop Temperature Adjustment

According to the table below, the 75°C column, room temperature helps reduce the wire size. The correction factor for this range is 1.05, so: 115 *1.05 = 120.75 amps. The room temperature increases our selected wire's ampacity. We can now check if a 4-gauge wire (85 ampacity) can handle 100 amps or not:

Ambient Temperature	75°C-Rated Conductor	90°C-Rated Conductor
21-25	*1.05*	1.04
26-30	1.00	1.00
31-35	0.94	0.96
36-40	0.88	0.91
41-45	0.82	0.87
46-50	0.75	0.82
51-55	0.67	0.76
56-60	0.58	0.71
61-65	0.47	0.65
66-70	0.34	0.58
71-75		0.50
76-80		0.41

85 amps * 1.05=89.25amps

Unfortunately, we need to stick to the previously-selected 2-gauge wire.

Since the wires are located freely at room temperature, there is no need to worry about the consideration of the height of the conduit on the roof.

Step 3: Number of Wires in a Conduit

Since we have only two wires in a conduit, in a dry and cool condition, there is no need to worry about this step. If four or more wires are being used, then we have to check the NEC table 310.15 (B)(3)(a), and we can use the less expensive THW wires.

Conclusively, a 2-gauge THW wire is required; therefore, the calculated wire gauge in step 2 is the smallest appropriate wire for the battery output.

Inverter Output Cable

The only remaining wires to size are the AC wires coming out of the inverter. Let's go through the steps:

Step 1: Ampacity

To figure out the current coming out of the inverter, we need to know two parameters.

1. The inverter's wattage.

2. The AC voltage of the grid or the appliances used in the house.

So,

The maximum current (amps) = inverter's continuous power/AC voltage.

In this case, the grid's electricity voltage is assumed 120 volts.

Maximum current generated by the inverter = 2000 watt / 120 volt = 16.66 amps

As you can guess now, a 14-gauge wire is the smallest wire that can handle up to 20amps.

Step 2: Ambient Temperature Adjustment

As we run these wires at room temperature, we will multiply it by 1.05, as previously extracted from table 310.15(B)(16) for 21-25°C temperature range:

20 * 1.05 = 21 amps

A 14-gauge wire is the smallest wire size that can handle 21 amps at room temperature, which is bigger than 16.66 amps; however, if the inverter's datasheet recommends a larger gauge, like 10 or 12, you should comply with that.

Step 3: Number of Wires in Conduit

Since only three wires- hot, neutral, and ground- are running through the conduit in the room temperature, no consideration is needed regarding the number of wires and the temperature adjustments.

If four or more wires are used, you need to follow the NEC table 310.15 (B)(3)(a)to adjust the amperage and the wire gauge accordingly.

PV Module Connectors

When handling a large solar system, you need connectors to link an array of PV modules. They connect a string of panels either in series or in parallel. The connectors consist of male and female parts. Different types of PV connectors are available on the market.

The most common types of connectors include MC3, MC4, PV, and Tyco Solarlock. These connectors are available in t-joint, u-joint, x-joint, and y-joint shapes.

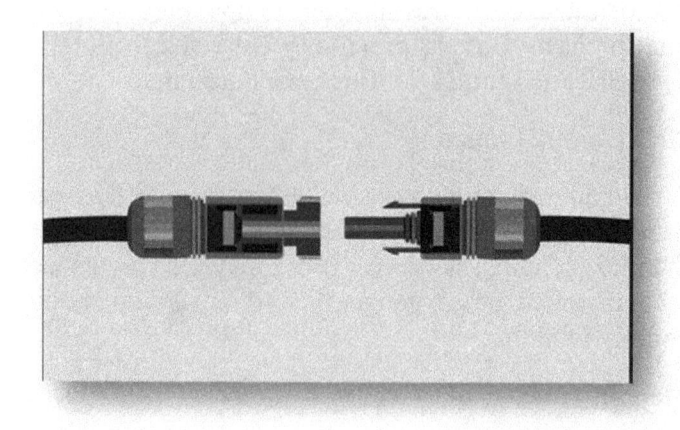

Wiring Solar Panels (Stringing)

Wiring/stringing of solar panels is a fundamental subject for every solar installer. You have to understand how different wiring configurations affect the voltage, current, and power generated by the PV array.

The panels are then connected to a load center to transfer the electricity to an inverter in a grid-tied system or a charge controller in battery-backed systems. Stringing can either be in series or parallel.

Series Connection

In a series connection, each PV module is connected to the next in a line. The positive terminal of one PV module is connected to the negative terminal of another PV module, and so on.

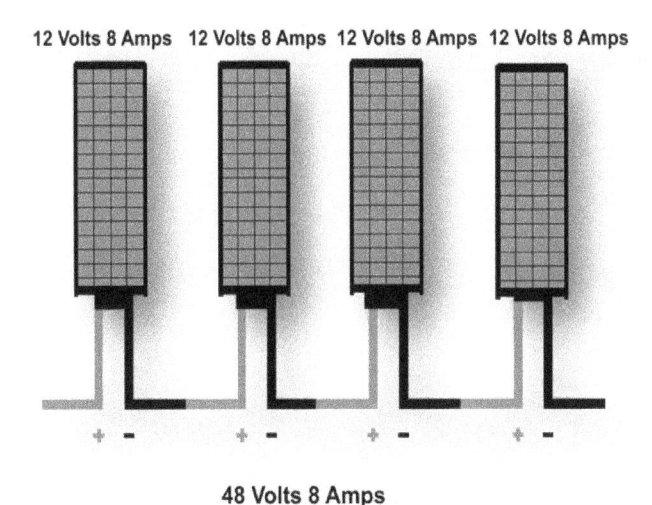

12 Volts 8 Amps 12 Volts 8 Amps 12 Volts 8 Amps 12 Volts 8 Amps

48 Volts 8 Amps

Each additional module adds up to the total voltage of the PV string while the current remains the same throughout the circuit. If you have four 12-volt modules rated at 8 amps, the total voltage in your PV string will be 48 volts while the current remains unchanged.

The power output would equal current multiplied by voltage. The only drawback of using a series configuration is that a shaded module significantly reduces the power output of the PV array.

Parallel Connection

In a parallel connection, the positive terminal of one module connects to the positive terminal of another module, while the negative terminals connect to each other. In this configuration, each additional module increases the current of the string; however, the voltage remains the same. And, as a result, shading in one module doesn't affect the power output of the entire PV string.

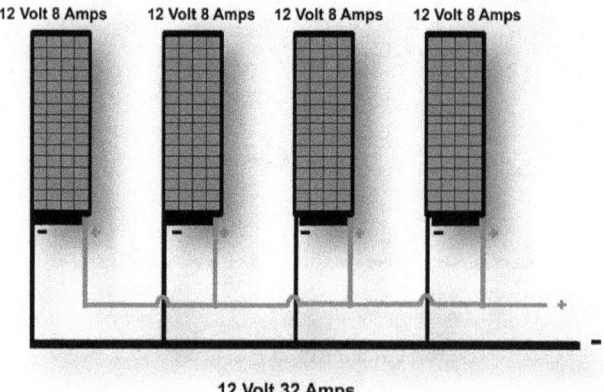

12 Volt 8 Amps   12 Volt 8 Amps   12 Volt 8 Amps   12 Volt 8 Amps

12 Volt 32 Amps

So having multiple modules in parallel will help generate power without exceeding the operating voltage of the inverter.

Positive and negative wires enter the combiner box on the rooftop in roof-mount systems. This box may function as a junction box or pass-through box, depending on the PV module configuration. The picture below shows a ground mount PV array; therefore, the combiner box is located under the panels near the ground.

Multiple PV modules connected in parallel join in the combiner box, then form a PV output circuit. The PV output then enters the load center located near other solar components like the battery and charge controller.

What to Know before Stringing the Panels

Your decision on whether to string the modules in parallel or series depends on your inverter, battery, and charge controller's specifications.

You have to check the inverter and charge controller's specification information found on the manufacturer datasheet to exactly match each other as well as your specific energy requirements. This information includes:

- The inverter's maximum input current.

- The inverter's maximum input voltage (V input, max), which is, in fact, the highest DC voltage the inverter receives.

- The inverter's minimum voltage (V input, min), which is defined as the minimum required voltage for the inverter to operate.

- Charge controller's maximum current and voltage.

- Battery's voltage.

PV Module Specification

In addition to the solar equipment information, you need to check this information about the selected PV modules:

- Open circuit voltage (Voc): this is the maximum amount of voltage the panels generate under no-load conditions.

- Short circuit current (Isc): this is the current that runs through PV cells when voltage is zero.

If you follow the exact steps as far as sizing your solar array, battery bank, charge controller and inverter you need to use all the above information to set up your system.

How to string your solar panels depends on your selected DC voltage (12, 24, 36, or 48) and the maximum amount of current you want your wires to carry. For instance, if your system consists of eight 100-watt, 12- volt panels with Isc of 8.5 amps, there are a few ways to string them depending on your desired battery, controller, and inverter's nominal voltage. The following diagrams show 12-, 24-, and 48-volt configurations for this PV system:

12 volt, 8.5 amps PV modules

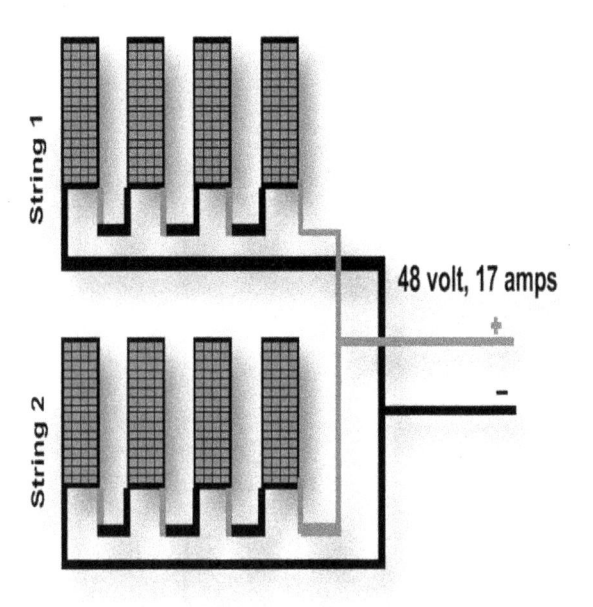

String 1

String 2

48 volt, 17 amps

+

−

The diagram above is more suitable for extensive PV systems with higher power output requirements. As you can obviously see, this type of wiring provides a fairly low output current, and consequently, smaller conductor diameters and lower expenses.

If you are looking for a medium-sized PV system, a 24-volt configuration is considered an appropriate option; however more parallel connections will lead to a higher amount of current generated by the PV modules when compared to the previous 48-volt configuration.

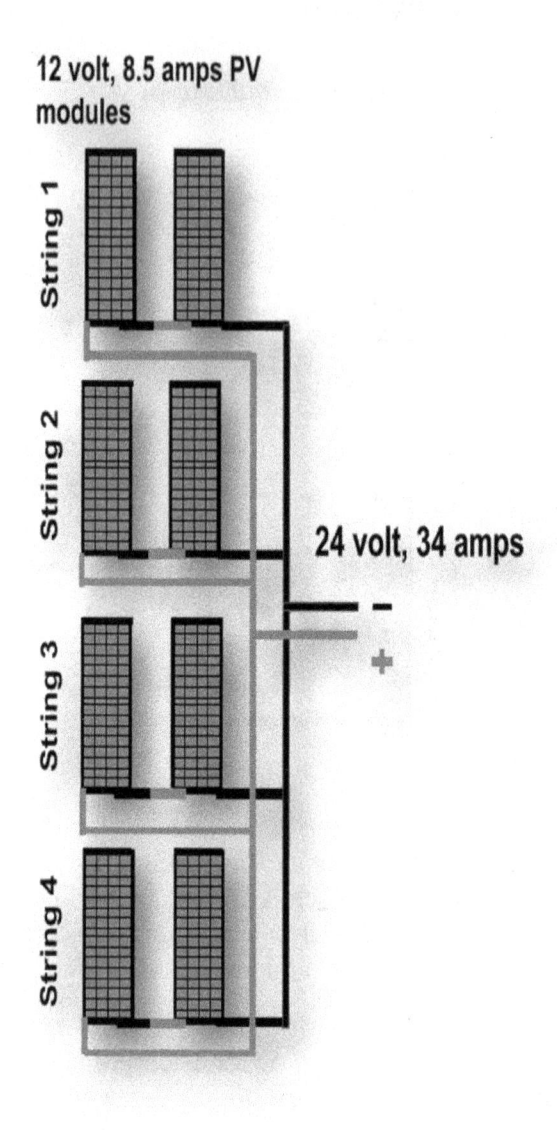

12 volt, 8.5 amps PV modules

String 1

String 2

24 volt, 34 amps

String 3

String 4

And finally, if you need a smaller PV system, the following parallel configuration can keep the voltage at 12; however, you may need to use fewer panels if you want to reduce the high amount of current output:

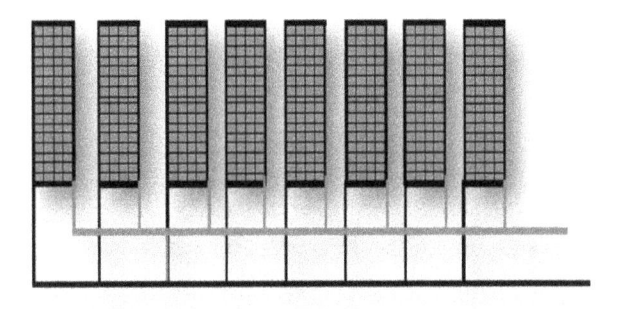

12 volt, 8.5 amps PV modules

Output: 12 volt, 68 amps

Chapter Summary

Choosing an appropriate wire size is essential for effective performance of your system. Sizing solar wires helps prevent overheating of the wires and reduces energy losses.

Avoid using wires that are not compliant with the recommended National Electrical Code; otherwise, your building inspector will not approve your installation.

Generally, the size of the solar wires depends on the amount of current generated by the panels and the length of the wire from the source to the electrical units. You can size your wires using the American Wire Gauge chart. Match your wire amp ratings with the wire

length to avoid voltage drop or power loss.

In the next chapter, you will see how protective devices will be added to your PV system to make it ready to start generating power.

Keep reading. You are almost there!

CHAPTER TEN: Connecting Overcurrent Protection Devices, Wrapping It Up, and Troubleshooting

Overcurrent protection (OCP) devices are crucial if you want to be prepared for unexpected events in any electrical system. They make the system run smoothly and safely by enabling the system to automatically turn off when an issue occurs. When it comes to connecting the PV system, the only accepted way to protect the wires and devices is to install OCP devices, which includes fuses and circuit breakers.

The OCP devices are required for each segment or circuit of the system to protect the wires. OCP devices protect the conductors from getting too hot and potentially, catching fire. The same applies to the electrical devices connected to the system- the OCP shuts off the power when a short circuit happens, thus keeping all the electrical devices safe.

Circuit breakers and fuses consist of an element that melts down and turns off the circuit when more than a certain amount of current (amps) passes through them. They should be sized smaller (in amps) than the wire they protect in order to block the current before the wires heat up. As mentioned, there are two major OCP devices frequently used for electrical wiring: circuit breakers and fuses. Circuit breakers have the following advantages over fuses:

1. Function as a disconnect switch, providing access for system repair.

2. Can be replaced or removed under electrical load.

3. Can be reset after the issue has been resolved.

4. Can be installed without a holder.

Circuit breakers work slower than fuses, so they would let the spikes happen. As spikes may harm the electrical devices in a mobile PV system such as a van, fuses are preferred to breakers in small-scale, off-grid PV systems.

Fuses are one-time-use devices, meaning that they must be replaced once an overcurrent event happens; however, they do function faster than a circuit breaker. They also need a fuse holder to be installed. Fuses have their own advantages:

1. Less expensive.

2. Available in higher voltages, more common in grid-tied systems.

DC and AC Load Centers

The load center is a place where the majority of the system's breakers and fuses are connected. Designating two separate DC and AC load centers provides numerous benefits. The main advantage is the ability to use the grid and the battery when the solar array should be switched off for any reason.

The DC load center is located in the same place where other solar components such as the battery, controller, and inverter are installed. It contains circuit breakers for the following circuits:

1. Between the PV output and the charge controller.
2. Between the controller and the battery.
3. Between the battery and the inverter.

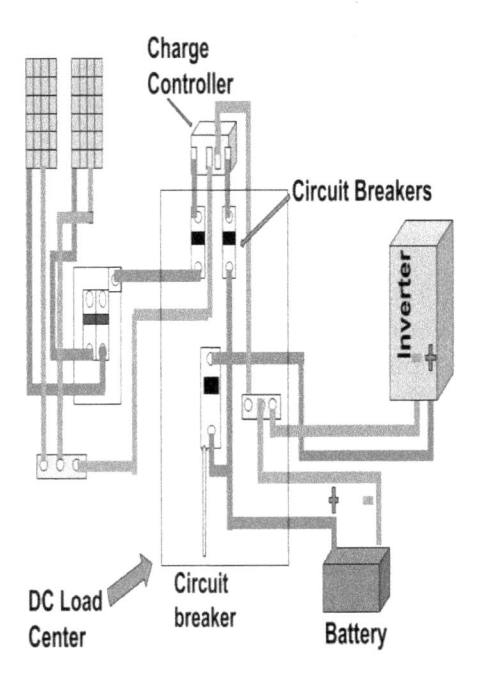

Basic Rules for OCP Device Sizing

Fuses and breakers must be connected as close to the battery (or "hot" end) as possible, so that there will be less length of wire that can still carry power in case the fuse switches off.

Additionally, breakers should always be sized smaller than the wire. Only in this way can they melt down and shut the power off before the wires heat up. For instance, for a 30-amp charge controller wired to a battery via a 40-amp rated wire, a 30-amp fuse is recommended.

Despite the mentioned differences, sizing circuit breakers and fuses follow the same rules. They are sized according to two electrical features:

1. Current in amps.

2. Voltage (DC and AC) in volts.

Maximum Current:

According to the NEC, we need to oversize the Isc of each segment of the DC circuit by 25% due to more than three hours of continuous use, exactly as you were required to do while sizing your system's wires.

Only up to the charge controller, we need to add another 25% due to over irradiance as well. This additional current may be generated when the irradiance goes higher than the Standard Test Condition (STC) in sunny days. Past charge controller, over-irradiance correction is not needed.

Voltage:

You need to match the voltage of your system (battery, inverter, and the array) with the breaker or fuse. OCP devices are also rated for DC and AC. Do not forget to install the DC and AC breakers for the corresponding segments of the system.

Four locations where installing an OCP device is almost always recommended are as follows:

1. The combiner box.

2. Between the solar array and the charge controller.

3. Between the charge controller and the battery.

4. Between the battery and the inverter.

Three of these OCP devices will be connected in the load center while the other one is located in the combiner box on the roof. Let's size the proper circuit breaker size for each of the PV system's circuits:

Combiner Box

For each string of panels, one breaker should be installed in the combiner box on the roof; therefore, the breaker should be able to handle the maximum current of each string of panels. So the correct size is calculated as follows:

Isc of the string * 125% (3 hour use) * 125% (over irradiance)

For example, if we have eight 115-watt, 12-volt panels connected in two parallel strings, with 4.8 amps Isc, the output of each string would be 48 volts, and still 4.8 amps Isc because all the panels in a string are connected in series, which means that the current remains constant.

Therefore, the maximum current = 4.8 * 156% = 7.48

A 10-amp fuse or breaker can handle this amount of current; however, an important point to remember is that the OCP size should always be smaller than the wires they need to protect. So in this case, the wire must be capable of handling at least 15 amps.

Between the Solar Array and Charge Controller

The size of the OCP device between the solar array and charge controller depends on the current generated by your solar array. As we have covered the details about the maximum current (Isc) in previous chapters, you can easily calculate your array's short circuit current (Isc) no matter how you have connected your solar panels: series, parallel, or both.

Since you have four 115-watt, 12-volt panels, with 4.8 amps Isc connected in series in each string, the output of each string would be 48 volts and still 4.8 amps because in series connection, the current remains constant. Since this breaker should protect a wire that carries the current of parallel strings, you should consider the number of strings as well.

As previously explained, two different safety factors must be considered for the PV arrays' generated current: add 25% for over-irradiance (extreme summer noon sunlight) and another 25 % for more than three hours of continuous use. Therefore, to calculate the right wire size, we need to multiply the Isc of each string by the number of strings as well as by 1.56.

Maximum current = Isc * number of parallel strings * 1.25 * 1.25

Max current = 4.8 * 2 * 1.56 = 14.96 amps.

In this case, a 15-amp fuse or breaker is the appropriate size to protect a wire with at least 20 amps of current-carrying capacity.

Between the Charge Controller and Battery

The OCP size needed between the charge controller and battery bank considers the maximum amperage rating of the charge controller. The breaker selected must be equal to or greater than the controller's amp rating, as well as smaller than the wire used. In other words, the breaker should always be sized smaller than the wire.

The alternative way to size this circuit breaker is to multiply the Isc of the string by the number of strings by 1.25 (safety factor):

Max current (amps) = Isc (amps) * number of parallel strings * 1.25 (safety factor).

Between the Battery and the Inverter

The fourth OCP device recommended is supposed to protect the inverter's wire located between the battery bank and the inverter.

Most inverter manufacturers recommend the fuse size or equip it with a built-in breaker or fuse inside the inverter; otherwise, you should divide the maximum continuous power of the inverter by the nominal battery voltage and add 25% to it for the safety factor to calculate the maximum current possible. For instance, for a 5000-watt, 48-volt inverter connected to a 48-volt battery, we will have the equation below:

$I = P/V$

I (amps) = 5000 W / 48 V * 1.25 = 130 amps

So a 150-amp fuse or breaker is recommended.

How to Crimp

To join electrical circuits, electrical connectors are used, and this is done by crimping and attaching the connector to the wire. Although this may often be neglected, the quality of the connection depends on the quality of the crimping of the wire attached to the connector.

You can learn how to crimp step-by-step below:

Step 1: Select Your Crimping Tool

A ratchet crimper is more expensive as well as more durable, and requires less effort. The crimping die on the crimper must match the wire gauge used.

So, what is a crimping die? It is the piece at the top of the crimper that actually crimps. All crimpers have one.

Step 2: Stripping

Place a quarter inch of the wire into the die. Locate the hole associated with the wire's gauge in the crimping die. If crimping a 14-gauge wire is intended, for instance, the hole designed for a range covering this gauge should be selected. Crimping dies usually have color-coded holes designated for certain wires.

Apply pressure to the tool's handles, pull it away from the wire, and remove the plastic insulation. Now you should have a quarter-inch of the wire stripped.

Step 3: Applying Force

A manual crimper needs the jaws of the crimper to be held perpendicular to the wire. This process takes some practice to do perfectly. Twisting the exposed wire, make it more firm and compact. This will result in a more condensed wire and consequently, a stronger connection to the connector.

Insert the wire into the connector (wire lug) so that the insulation touches the barrel. The wire should not pass the barrel more than a quarter of an inch. Now you should place the connector's barrel into the designated crimping slot. Placing the wire horizontally with the barrel facing upward is preferred.

Some tools have colored markings for AC wires, so by matching the insulation color with the associated color, you can select the right size; otherwise, or if you are crimping DC wires, just match the wires with the gauge markings on the side of the tool.

Now you should hold the tool steady and vertically toward the floor and squeeze its handle over the wire lug's barrel with force. When using manual crimpers, squeeze the crimper's handles with a considerable amount of force to secure a durable, high-quality connection. Since most of the wires being crimped for solar purposes are thick enough, there is not high chance of over-crimping; therefore, apply as much pressure as you can, especially when using a manual crimping tool.

Ratchet crimpers, however, will be aligned over the lug's barrel automatically by its crimping slot. Less effort is needed when using a ratchet crimper. You can now tug the wires apart and see how strong the connection is. They should not tear apart even when applying considerable force.

Sometimes, the connection is still loose because of improper wire or wire lug size. Soldering the joint between the lug and wire is an option; however, it is not recommended due to the fact that the solder alloy is different from pure copper and this will result in future issues. Redoing the crimp is the most reliable option.

Step 4: Heat Shrinking

The last step is to seal the crimped wire and wire lug with a heat shrink. Always seal the terminal when your connection is going to be exposed to heat, snow, wind, or rain.

Final Checklist and Troubleshooting

Home PV systems are unique and have become an outstanding investment for individuals looking for a clean source of energy. Solar power installation provides a number of benefits, not only in residential homes, but also for commercial properties.

Generally, solar power systems require low maintenance and have a payback period of between 5–7 years; however, there are some problems associated with solar systems, and if appropriate actions are not taken early, the problems will worsen; therefore, it is necessary to recognize the major issues you may experience so that you can take appropriate action if needed.

Issues to Look Out For

Sometimes, identifying issues with your solar system might seem challenging. For example, you may notice the system is not producing as much power as before, and you don't know why. Changes in PV system performance are often a result of one or more of the following reasons:

1. Loose Wiring and Connections

The PV system includes a wiring network that links various PV modules together and connects them to the solar inverter and the battery. If the wires connecting to the solar components become loose (almost always due to poor crimping of the connectors), it will affect the overall performance of your solar system.

You can use an ammeter and a voltmeter to identify wiring faults. Once you identify where the fault is, you can take action to avoid further damage to your system. You can also talk to an expert to help you solve the problem. If you have already installed a monitoring system that allows you to monitor the performance of the panels via your smartphone, it makes your troubleshooting work much easier.

2. Overheating of the System

If your array experiences extreme temperatures, it may affect the performance of the panels. Although this is not an indication that the panels are faulty, you may notice less power production during hot days.

Sometimes, high temperatures can affect some sections of your solar panels, and this makes them wear down much faster.

If you live in an area that experiences high temperatures, you had better purchase solar panels that perform better under high-heat areas. For example, Panasonic HIT panels have thin layers that protect the monocrystalline layer, making the panels more efficient in high-temperature areas.

3. Solar system Not Performing as Expected

Sometimes solar panels don't work as expected due to a number of factors. If you continue to receive low current and low voltage, it may be from one of several possible issues, so, you have to perform the right diagnostic steps to correct the issue. Always record information about the error messages, when you first noticed the problem and the number of times the problem occurs after you first noticed the problem. With this information, you can easily troubleshoot your panels or hire an expert to help you fix the issue.

4. Your System is Dirty

Dust, debris, and dirt can affect the efficiency of the panels. Though these environmental issues may appear minor, they can reduce the amount of power generated by the panels. To avoid this, you need to regularly clean or remove all the dirt from the panels to avoid shading.

Common Mistakes Solar Installers Make

The switch to a solar power system is gaining popularity in many US homes. If you're considering switching to solar energy, there are a number of mistakes you need to avoid if you want to obtain the best out of your investment. Some of these common mistakes include:

1. Not Taking into Account the Structure of Your Home Roof

Before mounting solar panels on your roof, you have to consider if your roof is in good enough condition to withstand the weight of the panels; therefore, you have to make sure that the structure of your roof can support the weight from the panels.

Does your roof have chimneys, swamp coolers, or air vents? These structures make it difficult to install the panels near the main unit. That is why it is worth considering your roof type before installing the panels. For example, if your rooftop is made of fine cement that makes the roof brittle, then you should reconsider your racking system.

2. Improper Sizing of the Solar Panel System

When sizing your solar system, most of the time you focus on calculating the total load and installing panels that help meet your energy demands. You may forget to consider other factors such as voltage setting of the inverter, panel orientation, shading, climate, and efficiency drop that affect the amount of solar power generated by the panels.

Unless you account for these factors, you will not be able to choose a properly-sized array.

If your solar array and battery bank don't have the same voltage, it will affect your entire system's performance.

3. Buying the Cheapest Solar Panels

Most of the time, you switch to solar power to reduce utility bills. For this reason, you may go for the cheapest solar panels, particularly if you're on a budget. Though buying the panels at a lower cost may seem ideal for now, it may end up being expensive in the long run.

Sometimes, you might end up replacing your entire PV system after a while because it is faulty. For this reason, you shouldn't purchase panels because they're cheap. You should evaluate the designs and prices of different types of panels before choosing the best panels for your home.

4. Mismatching Battery Sizing

Matching your battery bank with the right charge controller and the appropriate wire size is particularly important for the proper functioning of your charging source.

When sizing an off-grid system, your solar array must generate enough power to keep the battery fully charged, and at the same time not overcharge the batteries. Overcharging and over-discharging affect the battery life.

5. Not Planning Ahead

When planning to go solar, people often tend to only consider their present energy needs. Not everyone expects their energy needs will change in the future. When designing your PV system, it's wise to think about possible future changes in your energy demands. For example, if you're planning to buy an electric vehicle, ask yourself whether your solar system is expandable? And how many panels can fit on your rooftop? Since most solar panels can last up to 25 years, you can plan ahead in case you want to expand your system and don't have enough space to install the panels on the rooftop.

If you add more panels to the system, you should also size the solar inverter, charge controller and battery bank to match your expanded array. As inverters have their own power limitation, they may not support a larger PV system. So, expanding your system is not considered as just adding panels to the existing ones; it is best that all other components are oversized in advance if you anticipate higher energy consumption in the future.

Chapter Summary

In this chapter, you learned where to connect overcurrent protection devices within your PV wiring network. You also now know how to size OCP devices. Additionally, you've seen some of the problems you may experience with your solar system installation, how to troubleshoot them, and how to avoid the common mistakes other solar installers often make.

If you are confident enough to switch to solar, now it's time to take action. If not, you may need to review some more extensive procedures, such as PV module sizing and wire sizing. Alternatively, you may need to consult a solar specialist.

Final Words

Solar energy is among the most easily accessible forms of renewable energy. And as a result, more people are welcoming solar power as their alternative source of energy. The main focus of this book is to help you learn how to harness solar power in your homes in a DIY manner; however, before you can start installing a solar system, you must have a fundamental understanding of how electricity works.

This understanding will help you appreciate how the system's components interact and work together, thereby giving you invaluable data for troubleshooting your system. This final section is meant to offer you an overall review on electricity and photovoltaic systems.

Electrical voltage, measured in volts (V), represents the electromotive force (E). This force is often referred to as potential difference because there should be a difference between the power source voltage and the load voltage in order to create a current.

Electrical current can be defined as the rate of charge flow. There are two types of current that flow through a circuit: direct and alternating current. Direct current (DC) is a constant current that flows in one direction, moving from a high voltage point (positive) to a low voltage point (negative).

The main disadvantage of using DC electricity is that it limits the distance in which current is transmitted. This makes DC transmission for shorter distances greater than transmitting for longer distances.

Alternating current (AC) describes the flow of electrical charge in different directions; that is, the change in voltage and current level. As the current keeps on switching direction, it creates a frequency, which is measured in hertz (Hz). The faster the switching is, the higher the frequency will be.

In an electrical circuit, resistance affects the flow of current. If there is high resistance in the material, there will be little or no current flowing through it. These materials are referred to as insulators. Conductors have low resistance and allow electrons to flow freely. The conductor acts as a highway while the electrons represent the cars. The more lanes the highway has, the more cars can travel, and vice-versa.

Therefore, it is important to determine circuit resistance when installing your system. The simplest way is to use Ohm's law to determine the circuit's resistance by calculating the total voltage drop across the circuit.

Once you know the basic electrical components, you will need to know the different tools and solar installation components required to set up your PV system. To start with, you have to consider safety precautions and protective gear for installing the panels. Other items include power tools for solar system installations, mounting tools, wiring tools and battery installation tools.

Solar systems are designed in three ways; grid-tied systems, hybrid systems, and off-grid systems. Grid-tied systems are connected to an electrical grid to draw power during low peak sun hours. During peak hours, you can send the excess power to the utility company. Electricity on the grid is transmitted as AC, and most devices are calibrated to use AC power.

If you want to access power during a power outage in the electrical grid, you can add a battery to your system. This hybrid system allows you to store excess power in the battery for use on those days where no power is generated. This system also makes you less reliant on grid power.

Lastly, you can decide to go off-grid by designing an on-site solar power system. An off-grid solar system allows you to be energy-independent by designing solar systems based on your energy needs.

When designing your solar system, you should consider factors such as shading, efficiency drop, panel orientation, and climate. All these things can affect the output of the system.

The efficiency of the solar system drops each year, and it is important to account for that when sizing your solar panels. In addition, you have to look at the panel's Voc temperature coefficient and its impact on the performance of the panels.

Solar panels are usually tested under ideal conditions, but in the real world, your panels are exposed to extreme weather conditions. For example, high temperatures reduce the efficiency of the panels as well as the amount of energy generated.

Your location also dictates the peak sun hours your panels are exposed to each day. During peak hours, your panels generate more power. Most areas experience 3–6 peak sun hours.

Further, your system needs to be designed based on the voltage of the solar components. If you don't select the right voltage for your panels or the battery bank, it will affect the overall performance of the system. Seasonal temperature changes are also another factor that affects the voltage of your setup.

Depending on your energy needs, you can size your solar panels and come up with the correct number of panels you need to meet your energy demands for your home. When sizing the panels and battery, you have to consider your daily consumption rate, days of autonomy, and any other factor that might affect the actual performance of your PV system.

After sizing your PV system, you can go ahead and size your solar controller, inverter, and battery bank. If you're installing an off-grid system, you need to calculate your daily power consumption to help estimate the number of solar panels you need to meet your energy needs.

You have to size your roof to know whether the number of panels you need can fit your rooftop space. Come up with the design layout of your roof and install the panels. If you don't have enough rooftop space, you can decide to use a ground mount solar panel installation or purchase more efficient PV modules and PV components.

When mounting panels on a sloped rooftop, you can secure them either through a rail system, rail-less system, or shared-rail system; otherwise, you can use a flat roof mounting system if you have a flat rooftop.

Follow the panel installation process steps explained earlier to install your solar panels on your rooftop or ground mounts. Once you install the panels, go ahead and install a solar controller, which acts as a regulator to control the amount of power going into the battery bank from the PV array. The charge controller charges your batteries based on the set level so as to maintain the battery life.

The charge controller uses two technologies: Pulse Width Modulation (PWM) and Maximum Power Point Tracking (MPPT). The controller automatically turns off the running load when the battery is low and turns it on after the battery is charged. The size of your charge controller depends on the current generated by your PV array and your system voltage.

When selecting the battery bank for your system, you have to consider battery lifespan, power capacity, along with depth of discharge, cycle life, and battery voltage. All these factors will help you choose the best battery bank that meets your power needs for a number of days. And, depending on the type of PV system you want to install into your home, you can size your solar battery to meet your energy demands.

Lastly, you have to choose the right type of wires, connectors, and protective devices to maximize your power output and prevent fire hazards that could result from faulty wiring. Always ensure you have the right wire gauge that can handle the amount of current generated by the panels. You also have to ensure the wires are rated at the correct ampacity and take into account the wire length.

There are different types of PV module connectors that allow you to wire the panels either in series, parallel, or both. Once you connect all the solar components, you can go ahead and test if your system is functioning properly. If you experience some challenges, you can troubleshoot your solar system and correct any mistake you have made in the installation process. Continuously monitor the system to ensure all the components are working correctly.

DIY SOURCE BOOKS created this book to provide a practical, step-by-step manual to ease the complicated process of designing and installing a PV system for those who want to go solar. Since all the aspects of the powering procedure is covered in detail, some parts of the book such as panel sizing and wire sizing, may seem overwhelming; however, this book has clarified all of the topics in a stepwise manner. As such, you may need to review some chapters a second time before starting your solar design and installation.

We wish you best of luck with your solar project!

References

AltEstore. (n.d.) "Solar power components, part 4." Youtube video, 10.19.
https://www.youtube.com/watch?v=ZdKiO2zLF5g&t=498s
Are solar panels worth it? (n.d.) Unboundsolar.com.
https://unboundsolar.com/solar-information/return-on-solar-investment
Arthur. August 14, 2019. Eight factors to consider before installing solar panels. Ledwatcher.com.
https://www.ledwatcher.com/8-factors-to-consider-before-installing-solar-panels/
Battery bank sizing. (n.d.) Alternative energy.org.
https://www.altenergy.org/renewables/solar/DIY/battery-bank-sizing.html
Battle born batteries. (n.d.)
https://batttlebornbatteries.com/product/12v-lifepo4-deep-cycle-battery/
Beaudet, Amy. April 13, 2016. How to read solar panel specifications. Altestore.com.
https://www.altestore.com/blog/2016/04/how-do-i-read-specifications-of-my-solar-panel/
Brown, Gwen. October 15, 2019. Solar panel wiring basics. Solar power world.
https://www.solarpowerworldonline.com/2019/10/solar-panel-wiring-basics-an-intro-to-how-to-string-solar-panels-2/
Chaaban, Mohammed. (n.d.) Irradiance and PV performance optimization. E-education.psu. https://www.e-education.psu.edu/ae868/node/877
Chaaban, Mohammed. (n.d.) Wiring within a PV module and the shading effect. E-education.psu. https://www.e-education.psu.edu/ae868/node/875
Choosing the correct wire size for a DC circuit, part 1. (n.d.) Blue Sea.com.
https://www.bluesea.com/resources/1437
Choosing the right solar charge controller. (n.d.) Solar for rvs.com.

https://www.solar4rvs.com.au/buying/buyer-guides/choosing-the-right-solar-charge-controller-regulat/

Choosing the right wire size. (n.d.) Windy nation.com. https://www.windynation.com/jzv/inf/choosing-right-wire-size

Clifton, Stephen. (n.d.) How to fuse your solar system. Renology.com. https://www.renogy.com/blog/how-to-fuse-your-solar-system/

Conductor wire sizing. (n.d.) Penn state. E-education.psu. https://www.e-education.psu.edu/ae868/node/966

Das, Santosh. February 1, 2020. Solar panel installation guide. Electronics and You.com. http://www.electronicsandyou.com/solar-panel-installation-guide-step-by-step-process.html

De Rooij, Dricus. (n.d.) Solar panel angle. Sino Voltaics.com. https://sinovoltaics.com/learning-center/system-design/solar-panel-angle-tilt-calculation/

Different types of solar systems. (n.d.). Solar quotes.com. https://www.solarquotes.com.au/good-solar-guide/system-types/

DIY off-grid solar system. (n.d.) Instructables workshop. https://www.instructables.com/DIY-OFF-GRID-SOLAR-SYSTEM/

DIY Solar Jon. (n.d.) How to size your off-grid solar batteries. Instructables workshop. https://www.instructables.com/How-to-Size-Your-Off-Grid-Solar-Batteries-1/

Everything you want to know about solar wires and cables. March 8, 2021. Central Plain Cables and Wires Co.https://www.zw-cable.com/news/Everything_You_Need_to_Know_About_Solar_Wires_and_Cables.html

Five Battery types explained. (n.d.) Battery stuff.com https://www.batterystuff.com/kb/articles/5-battery-types-explained-sealed-agm-gel.html

Grid connected PV system. (n.d.). Alternative Energy tutorials. https://www.solarreviews.com/blog/grid-tied-off-grid-and-hybrid-solar-systems

Grid-tied, off-grid, and hybrid solar systems. July 15, 2019. Solar Reviews.com. https://www.alternative-energy-tutorials.com/solar-power/grid-connected-pv-ystem.html

Guide to understanding battery specifications. December 2018. MT electric vehicle team. pdf. http://web.mit.edu/evt/summary_battery_specifications.pdf

How do solar batteries work? October 7, 2019. Infinite energy.com. https://www.infiniteenergy.com.au/how-do-solar-batteries-work/

How do solar batteries work? Updated July 6, 2021. Energy sage.com. https://www.energysage.com/solar/solar-energy-storage/how-do-solar-batteries-work/

How does a grid-tiered solar system work? February 25, 2019. Freedom Forever.com. https://freedomforever.com/how-grid-tied-solar-system/

How does a lithium-ion battery work? September 14, 2017. Energy.gov. https://www.energy.gov/eere/articles/how-does-lithium-ion-battery-work

How does temperature affect battery performance? (n.d.) CED Greentech.com. https://www.cedgreentech.com/article/how-does-temperature-affect-battery-performance

How many solar panels will fit on your roof? Updated February 28, 2020. The solar nerd.com. https://www.thesolarnerd.com/blog/how-many-solar-panels-fit-on-roof/

How to crimp a wire. May 13, 2020. Wikihow.com. https://www.wikihow.com/Crimp-a-Wire

How to design solar PV system. (n.d.) Leonics.com. http://www.leonics.com/support/article2_12j/articles2_12j_en.php

How to size an off-grid solar system. December 13, 2018. Unbound solar.com. https://unboundsolar.com/blog/sizing-off-grid-solar-system

How to size a solar system. July 14, 2020. Unbound solar.com. https://unboundsolar.com/blog/how-to-size-solar-system

Hybrid inverter comparison chart. (n.d.) Clean energy reviews. https://www.cleanenergyreviews.info/hybrid-all-in-one-solar-inverter-review-mpp

Hyder, Zeeshan. May 24, 2021. Three easy ways to calculate solar system size. Solar reviews.com.

https://www.solarreviews.com/blog/what-is-the-right-size-solar-array-for-my-home

Introduction to Solar Power System. January 15, 2021. Solar smiths.com. https://www.solarsmiths.com/blog/knowledge/introduction-to-solar-power-system/

Inverter basics. (n.d.) Wind and sun. Solar electric.com. https://www.solar-electric.com/learning-center/inverter-basics-selection.html/

Kunz, Natalie. (n.d.) Solar panel installation and maintenance. Green match.com. https://www.greenmatch.co.uk/blog/2014/09/solar-panel-installation-and-maintenance

Lane, Catherine. January 10, 2021. What are the different types of solar batteries? Solar reviews.com. https://www.solarreviews.com/blog/types-of-solar-batteries

Lasswell, Robert. April 4, 2021. Solar Panel Installation. Semprius.com. https://www.semprius.com/solar-panel-installation/

Making sense of solar panel specifications. (Update March 22, 2021) The solar nerd.com. https://www.thesolarnerd.com/guide/understanding-solar-panel-specs/

Marsh, Jacob, August 29, 2019. AC versus Solar Battery coupling. Energy sage.com. https://news.energysage.com/ac-vs-dc-solar-battery-coupling-what-you-need-to-know

Niclas. (n.d.) Temperature considerations for solar batteries. Sino voltaics.com https://sinovoltaics.com/learning-center/storage/temperature-considerations-for-solar-batteries/#:~:text=A%20lead%20acid%20battery%20may,maintain%20a%20given%20charging%20current.

Off-grid solar systems: introductory guide. December 25, 2020. Solar reviews.com. https://www.solarreviews.com/blog/off-grid-solar-systems

Pradeepan, Nani. August 25, 2016. Choosing an effective solar PV system. New dawn energy solutions. https://newdawn-es.com/choosing-an-effective-solar-pv-system-for-your-need/?gclid=Cj0KCQiAnKeCBhDPARIsAFDTLTJeMrdBas cSPH17f2-XYoVne2hyGNnqxPZ1Sl3BOIzRzX47kofDN-UaAs5VEALw_wcB

Pros and cons of going off the grid. (n.d.) Carbon track.com. https://www.carbontrack.com.au/blog/off-grid-pros-cons/

Roger. October 24, 2018. How to increase the solar battery lifespan? Solar electric power company. https://www.sepco-solarlighting.com/blog/how-to-increase-the-solar-battery-lifespan

Rollett, Catherine. June 3rd, 2020. Solar costs have fallen 82% since 2010. PV magazine.com. https://www.pv-magazine.com/2020/06/03/solar-costs-have-fallen-82-since-2010/

Rosenblat, Lazar. (n.d.) Residential off the grid systems. Solar.smps. https://solar.smps.us/off-grid.html

Rushworth, John. January 9, 2019. Lead-acid battery charging in cold weather. Victron Energy.com. https://www.victronenergy.com/blog/2019/01/09/lead-acid-battery-charging-in-cold-weather/

Say hello to the next generation. November 17, 2017. Unbound solar.com. https://unboundsolar.com/blog/era-energy-excellence-hdwave-inverter

Series and parallel circuits. (last edited June 14, 2021) Wikipedia.org. https://en.wikipedia.org/wiki/Series_and_parallel_circuits

Shamieh, Cathleen. (n.d.) Closed, Open, and Short Circuits. Dummies.com. stps://www.dummies.com/programming/electronics/components/closed-open-and-short-circuits/

Solar array tilt angle. (n.d.) Ced greentech.com. https://www.cedgreentech.com/article/solar-array-tilt-angle-and-energy-output

Solar battery care, maintenance, and safety. (n.d.) Solartown.com. https://solartown.com/learning/solar-panels/solar-battery-care-maintenance-and-safety-dont-touch-the-terminals/

Solar cell I-V characteristic. (n.d.) Alternative energy tutorials.com. https://www.alternative-energy-tutorials.com/photovoltaics/solar-cell-i-v-characteristic.html

Solar charge controller basics. (n.d.) Solar electric.com. https://www.solar-electric.com/learning-center/solar-charge-controller-basics.html/

Solar electric photovoltaic modules. (n.d.) Solar direct.com. https://www.solardirect.com/archives/pv/pvlist/pvlist.htm

Solar electric system sizing. (n.d.) Solar Direct.com. https://www.solardirect.com/archives/pv/systems/gts/gts-sizing-sun-hours.html

Solar modules: why do tilt and orientation matter? (n.d.). Solar island energy. https://solarisland.energy/2019/04/solar-modules-why-do-tilt-and-orientation-matter/

Solar panel connectors and cables. (n.d.) Wind and sun. Solar electric.com. https://www.solar-electric.com/learning-center/how-to-use-mc4-connectors-cables.html/

Solar systems: how to calculate available roof space. October 18, 2017. Wind and solar.com. https://www.intermtnwindandsolar.com/solar-energy-systems-how-to-calculate-available-roof-space/

Storgaard, Morten. January 21, 2019. Can RV batteries freeze? Go downsize.com. https://www.godownsize.com/rv-battery-freeze/

Svarc, Jason. June 25, 2020. Solar battery system types. Clean energy reviews.com. https://www.cleanenergyreviews.info/blog/ac-coupling-vs-dc-coupling-solar-battery-storage

Svarc, Jason. August 10, 2020. What is a solar charge controller? Clean energy reviews.com https://www.cleanenergyreviews.info/blog/mppt-solar-charge-controllers

Tilt and azimuth angles. November 30, 2018. Unbound solar.com. https://unboundsolar.com/blog/solar-panel-azimuth-angle

Tools for a successful solar electric install. (n.d.) Alterstore.com. https://www.altestore.com/howto/tools-for-a-successful-solar-electric-install-a90/

Tools for DIY installation. (n.d.) Dynamic SLR. https://www.dynamicslr.com/tools-for-diy-solar-installation/

Tremblay, Sylvie. March 31, 2021. Ace your middle school science fair. Sciencing.com. https://sciencing.com/science-experiment-kits-for-middle-school-13763608.html

Types of circuits. (n.d.). Byjus.com. https://byjus.com/physics/types-of-circuits/

Understanding basic electrical theory. (n.d.) Library automation direct.com. https://library.automationdirect.com/basic-electrical-theory/

Voltage drop calculator. (n.d.) Unbound solar.com. https://unboundsolar.com/solar-information/voltage-drop
What does deep cycle mean? (n.d.) MKpowered.com. https://www.mkbattery.com/blog/what-does-deep-cycle-mean
What is a deep cycle battery? February 10, 2020. Relion battery.com. https://relionbattery.com/blog/what-is-a-deep-cycle-battery
What is a hybrid solar system? April 20, 2021. Solar Reviews.com. https://www.solarreviews.com/blog/hybrid-solar-systems
What is a solar charge controller? (n.d.) AltE store.com. https://www.altestore.com/store/info/solar-charge-controller/
Wire for solar panels. (n.d.) Alternative energy solutions. https://www.altenergy.org/renewables/solar/DIY/solar-wire.html
Wiring solar panels and batteries. (n.d.) Alterstore.com. https://www.altestore.com/diy-solar-resources/schematic-wiring-solar-panels-in-series-and-parallel/
Wiring your camper van electrical system. (n.d.) Parked in paradise.com. https://www.parkedinparadise.com/wiring/
Wiring your solar power system. (n.d.) AltE store. Youtube video, 21.22. https://www.youtube.com/watch?v=89u8R_aUFO4&t=795s
Image credits: Shutterstock.com
https://www.nfpa.org/NEC/About-the-NEC/Free-online-access-to-the-NEC-and-other-electrical-standards

DIY SOLAR POWER FOR BEGINNERS

CPSIA information can be obtained
at www.ICGtesting.com
Printed in the USA
BVHW051733030223
657818BV00005B/502

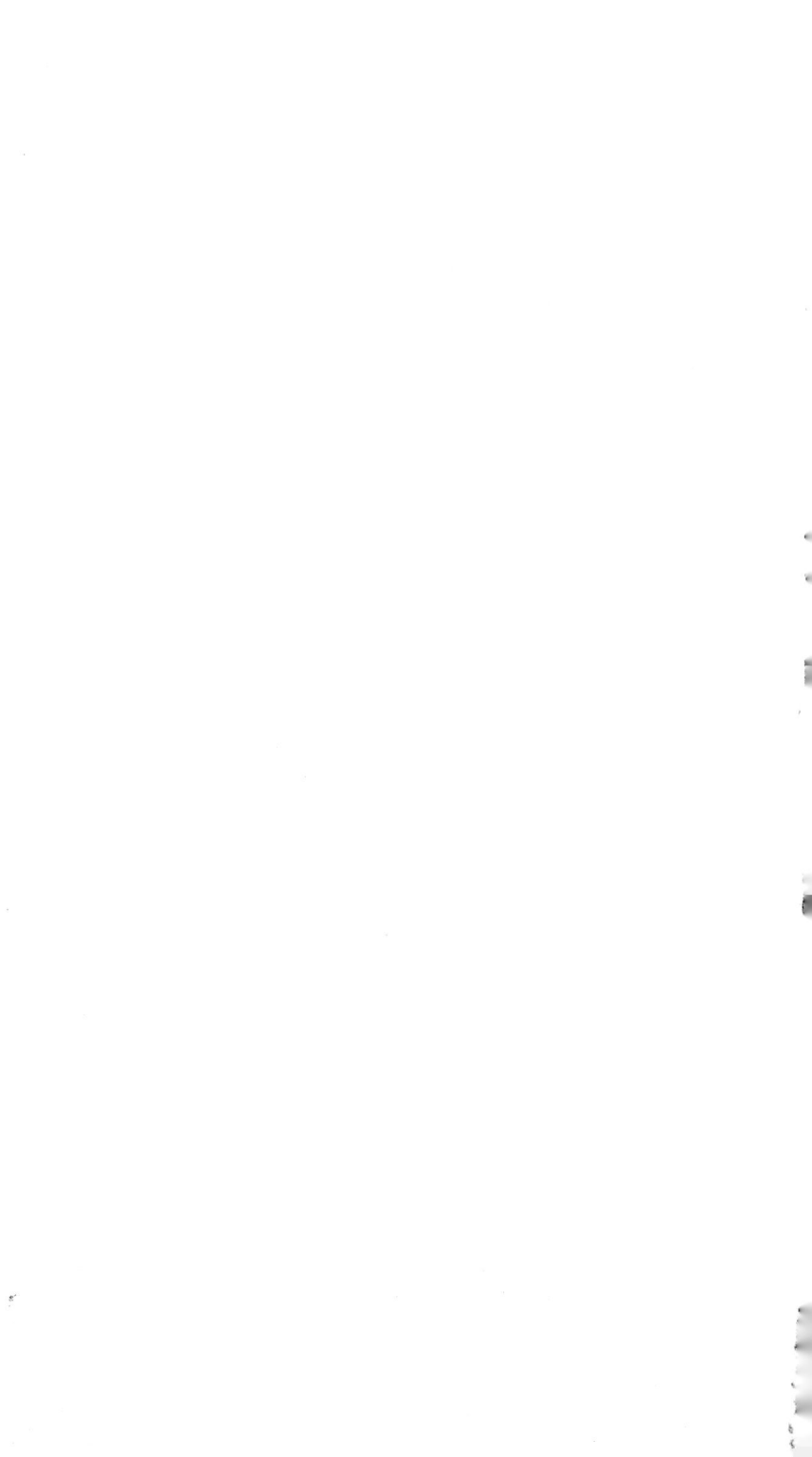